Rooms for the Learned Musician

A 20-YEAR RETROSPECTIVE ON THE ACOUSTICS OF MUSIC EDUCATION FACILITIES

RONSSE | LAWLESS | KANTER | CARREON BRADLEY

ACOUSTICAL SOCIETY OF AMERICA

On 27 December 1928 a group of scientists and engineers met at Bell Telephone Laboratories in New York City to discuss organizing a society dedicated to the field of acoustics. Plans developed rapidly, and the Acoustical Society of America (ASA) held its first meeting on 10–11 May 1929 with a charter membership of about 450. Today, ASA has a worldwide membership of about 7000.

The scope of this new society incorporated a broad range of technical areas that continues to be reflected in ASA's present-day endeavors. Today, ASA serves the interests of its members and the acoustics community in all branches of acoustics, both theoretical and applied. To achieve this goal, ASA has established Technical Committees charged with keeping abreast of the developments and needs of membership in specialized fields, as well as identifying new ones as they develop.

The Technical Committees include acoustical oceanography, animal bioacoustics, architectural acoustics, biomedical acoustics, engineering acoustics, musical acoustics, noise, physical acoustics, psychological and physiological acoustics, signal processing in acoustics, speech communication, structural acoustics and vibration, and underwater acoustics. This diversity is one of the Society's unique and strongest assets since it so strongly fosters and encourages cross-disciplinary learning, collaboration, and interactions.

ASA publications and meetings incorporate the diversity of these Technical Committees. In particular, publications play a major role in the Society. *The Journal of the Acoustical Society of America* (JASA) includes contributed papers and patent reviews. *JASA Express Letters* (JASA-EL) and *Proceedings of Meetings on Acoustics* (POMA) are online, open-access publications, offering rapid publication. Acoustics Today, published quarterly, is a popular open-access magazine. Other key features of ASA's publishing program include books, reprints of classic acoustics texts, and videos. ASA's biannual meetings offer opportunities for attendees to share information, with strong support throughout the career continuum, from students to retirees. Meetings incorporate many opportunities for professional and social interactions, and attendees find the personal contacts a rewarding experience. These experiences result in building a robust network of fellow scientists and engineers, many of whom become lifelong friends and colleagues.

From the Society's inception, members recognized the importance of developing acoustical standards with a focus on terminology, measurement procedures, and criteria for determining the effects of noise and vibration. The ASA Standards Program serves as the Secretariat for four American National Standards Institute Committees and provides administrative support for several international standards committees.

Throughout its history to present day, ASA's strength resides in attracting the interest and commitment of scholars devoted to promoting the knowledge and practical applications of acoustics. The unselfish activity of these individuals in the development of the Society is largely responsible for ASA's growth and present stature.

ROOMS for the LEARNED MUSICIAN

A 20-YEAR RETROSPECTIVE ON THE ACOUSTICS OF MUSIC EDUCATION FACILITIES

RONSSE | LAWLESS | KANTER | CARREON BRADLEY

Editors:

Lauren M. Ronsse, Ph.D.
Industry Fellow
Durham School of Architectural Engineering &
Construction
The University of Nebraska–Lincoln

Martin S. Lawless, Ph.D.
Visiting Assistant Professor of Mechanical
Engineering
The Cooper Union for the Advancement of Science
and Art

Shane J. Kanter
Senior Consultant
Threshold Acoustics

David T. Carreon Bradley, Ph.D.
Faculty Diversity Officer
California State University, Fullerton
Member of the Board of Trustees
Society for the Advancement of Chicanos/Latinos
and Native Americans in Science (SACNAS)

ISBN 978-3-030-72056-8
ISBN 978-3-030-72054-4 (eBook)
https://doi.org/10.1007/978-3-030-72054-4
© Springer Nature Switzerland AG 2021

Cover design by David T. Carreon Bradley

Primary Cover Photograph: original attributed to Myles Webb | modified based on www.flickr.com/photos/153417548@N02/40349332944

Background Cover Photograph: Kirkegaard Associates | Univeristy of Cincinnati | College Conservatory of Music - Werner Recital

This Springer imprint is published by the registered company Springer Nature Switzerland AG
The registered company address is: Gewerbestrasse 11, 6330 Cham, Switzerland

TABLE OF CONTENTS

xv   Editors' Preface

xxi   Acoustic Design of Music Education
        Facilities: An Overview

**Reflections from Key Design Team
Members**

3   The Soundscape of Music
        Rehearsal and Education Facilities
        by Gary W. Siebein

11   Architecture and Acoustics in an
        Era of Experimentation
        by Clifford Gayley

17   Audio/Video System Designers as
        Interpreters
        by Tim Perez

21   Supporting the Student in Theatre
        Design
        by Scott Crossfield

Music Education Facilities

25   List of Contributed Music Education Facilities

33   Primary & Secondary Education Facilities

87   Higher Education Facilities: Completed 2000-2014

169   Higher Education Facilities: Completed 2015-2020

247   Music Conservatories, Music Rehearsal Spaces, & Community Centers

Appendices

301   A: Glossary

311   B: Notes on Currency, Units, and Scale

315   C: References

Indices

321   A: Index by Space Type

325   B: Index by Location

331   C: Index by Acoustical Consulting Firm

Editors' Preface
RONSSE I LAWLESS I KANTER I CARREON BRADLEY

LAUREN M. RONSSE, PH.D.

Industry Fellow in the Durham School of Architectural Engineering & Construction, The University of Nebraska—Lincoln

Ronsse is an educator in architectural engineering and acoustics and has a Ph.D. in Engineering with a focus in Acoustics from the University of Nebraska—Lincoln. She also has experience in architectural acoustics consulting, including an enriching position as a Collaborating Consultant with Threshold Acoustics. Prior to her current Industry Fellow appointment, she was an Assistant Professor in the Audio Arts & Acoustics Department at Columbia College Chicago.

MARTIN S. LAWLESS, PH.D.

Visiting Assistant Professor of Mechanical Engineering, The Cooper Union for the Advancement of Science and Art

Lawless earned his Ph.D. in Acoustics in 2018 from the Pennsylvania State University where he investigated the brain's auditory and reward responses to room acoustics. At the Cooper Union, he continues studying sound perception, including 1) the generation of head-related transfer functions with machine-learning techniques, 2) musical therapeutic inventions for motor recovery after stroke, and 3) active noise control using an external microphone array.

SHANE J. KANTER

Senior Consultant, Threshold Acoustics

Kanter spends his days collaborating on the acoustic performance of his projects at Threshold Acoustics, and his nights and weekends going on adventures with his wife, son, and dog. Prior to arriving at Threshold, Shane received his Master of Arts in Architecture from the University of Kansas by studying with the venerable and beloved Bob Coffeen.

Rooms for the Learned Musician: A 20-Year Retrospective on the Acoustics of Music Education Facilities is the latest volume in a series of architectural acoustics compendiums published by the Acoustical Society of America (ASA). The previous two books in the series, which served as major inspiration for the current publication, are *Acoustical Design of Theatres for Drama Performance: 1985-2010*, published in 2010 and edited by David T. Bradley, Erica E. Ryherd, and Michelle C. Vigeant [1] and *Worship Space Acoustics: 3 Decades of Design*, published in 2016 and edited by David T. Bradley, Erica E. Ryherd, and Lauren M. Ronsse [2]. The first ASA book to focus on music educational facilities is *Acoustical Design of Music Education Facilities*, which was published in 1990 and edited by Edward McCue and Richard H. Talaske [3]. The present volume focuses on music education facilities—rooms for the learned musician—that were designed in the past twenty years and serves as a follow-up and update to the McCue and Talaske volume.

The current book is meant to be a valuable reference, resource, and inspiration to a wide audience, including acoustical designers and consultants, students studying music or building design, architects, music education facility directors, and musicians. The book features full-color spreads showcasing case-studies of 65 music education facilities from six countries across the globe. The architect or acoustical consultant for each facility contributed their project for inclusion in this book, which features 21 different acoustical consultants. The presentation of each facility includes a description of the space, photographs and/or computer-generated renderings, architectural drawings, and acoustical data. Most spaces are accompanied by a full-page architectural plan and section drawing. The descriptions and accompanying data showcase the architectural acoustics features, challenges, and highlights of the particular space.

There is a rich diversity of music education facilities covered in this book, ranging from music conservatories and community centers to primary, secondary, and higher education facilities. The book has been divided into four sections based on the type of facility: (1) Primary & Secondary Education (e.g., grade school and high school), (2) Higher Education (e.g. college and university) - completed from 2000 to 2014, (3) Higher Education - completed from 2015 to 2020, and (4) Music Conservatories, Music Rehearsal, & Community Centers. In cases where the facility type spanned more than one category, the venue was categorized according to the main facility type. For example, if a venue both serves higher education and is a conservatory, it typically appears in the Music Conservatory category.

Each facility has one featured space such as a recital hall or multi-use theatre. For ease of reference, the facility spreads in the book are color coded (see the table at the bottom right of this page) according to the type of featured space: Concert Halls - blue; Multi-Use Theatres - purple; Recital Halls - red; Rehearsal Rooms - green; and Other Spaces - orange. The featured space types are also listed in the information bar at the top of each spread.

In addition to the 65 music education facility case studies, the book also contains a series of essays intended to provide context to readers who may be new to the area of music education facility acoustics, as well as fresh insights for those experienced in the field. The first essay, by Gary Siebein (Professor Emeritus of Architecture and Acoustics at the University of Florida School of Architecture and Senior Principal Consultant with Siebein Associates, Inc.), provides historical context for the acoustical design of music education facilities, documents recent research in the field, and provides insights on the current status and future directions pertaining to the soundscape of music rehearsal and education facilities.

While the acoustical consultant is a critical member of the design team, the coordination and harmony of all design team members is necessary for the creation of a successful music education facility. The architect, audio/video systems designer, and theatrical consultant, to name a few, all play important roles in contributing to the final experience for the musicians and audiences. For this reason, essays from experienced professionals in these roles have also been included to discuss music education facility design issues from

DAVID T. CARREON BRADLEY, PH.D.
Faculty Diversity Officer, California State University, Fullerton

Carreon Bradley currently serves as the Faculty Diversity Officer at California State University, Fullerton. He previously served as the Vice President for Inclusion, Diversity, and Equity at Smith College, where he was also an Associate Professor in both the physics department and engineering program. Prior to that, he worked at Vassar College for over 10 years, where he was an Associate Professor of Physics and the chief diversity faculty-administrator. He is also a member of the Board of Trustees for SACNAS (Society for the Advancement of Chicanos/Latinos and Native Americans in Science). He conducts research on higher education organizational development, diversity and inclusion in higher education, access and equity in STEM, and acoustics - the latter for which he won the prestigious NSF CAREER Award.

Featured Space Type	Color
Concert Halls	
Multi-Use Theatres	
Recital Halls	
Rehearsal Rooms	
Other Spaces	

their unique perspective. Clifford Gayley (Principal at William Rawn Associates, Architects Inc.), discusses the combined responsibility of the entire design team when conceiving new spaces. He stresses the importance of meeting current and flexible needs of the community as demands continue to push toward interdisciplinarity. In the next essay, Tim Perez (Senior Consultant with Threshold Acoustics) comments on the audio/video systems design process. He presents an approach that augments the architect's vision for the project, while ensuring that the occupant and end-user needs are met. Finally, Scott Crossfield (Director of Design, Americas at Theatre Projects) pulls from his 25 years of experience as a theatre consultant to provide fundamental guidelines when designing music education facilities that prioritize and support the learning experience of the student. The music education facility case studies reflect the ideas and guidelines from these essays while featuring creative acoustic solutions designed to address some of the most difficult challenges associated with music education facility design.

For readers seeking more background information on architectural acoustics specifically related to music education facility design, an acoustical design overview with a general summary of the design process follows this Preface. The appendices include a glossary of key terms and a list of references for readers interested in architectural acoustics and music education facility design.

We are grateful to the many people who have helped this book come to fruition. In particular, we appreciate the work of the contributing architects and acoustical consultants whose firms' designs are featured in this publication. Their assistance was essential in compiling the necessary elements for each music education facility contribution, and they should be credited with much of this book's success. We are also indebted to our guest essay authors for their wonderful insights into the design of music education facilities. We also appreciate our colleagues in the world of acoustics who provided assistance throughout this process, especially Jonathan Weber, who served as a contributing editor for the book, and Michelle Vigeant, who was involved early in

the editorial process. We also thank University of Nebraska undergraduates, Christian Espinoza and Ben Ripa, who helped with managing the image assets for the book.

The process of creating this publication has been rewarding and challenging, and we hope that readers will find it to be a useful and valuable reference for many years to come.

Editors,

Lauren M. Ronsse, Ph.D.

Martin S. Lawless, Ph.D.

Shane J. Kanter

David T. Carreon Bradley, Ph.D.

References

[1] Bradley, David T., Ryherd, Erica E., and Vigeant, Michelle C. (Eds.). *Acoustical Design of Theatres for Drama Performance: 1985 – 2010*. Acoustical Society of America, 2010, New York.

[2] Bradley, David T., Ryherd, Erica E., and Ronsse, Lauren M. (Eds.). *Worship Space Acoustics: 3 Decades of Design*. Springer, 2016, New York.

[3] McCue, Edward and Talaske, Richard H. (Eds.). *Acoustical Design of Music Education Facilities*. Acoustical Society of America, 1990, New York.

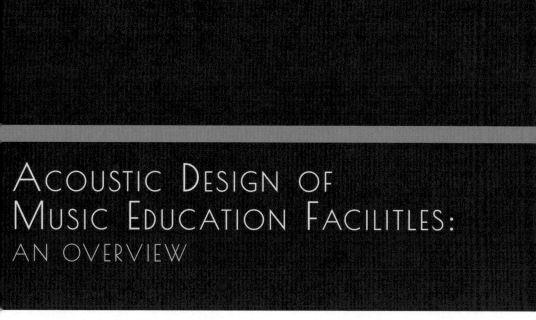

ACOUSTIC DESIGN OF
MUSIC EDUCATION FACILITLES:
AN OVERVIEW

A musician was walking home to his Manhattan apartment following a rehearsal. A tourist stopped him to ask, "Can you tell me how to get to Carnegie Hall?"

"Yes," answered the musician, "Practice!"

If you are lost in New York City trying to find Carnegie Hall, you should avoid the adverb "how" because you are only going to be faced with the timeless joke: practice, practice, practice [1]. For the learned musician, practice is the key to success. The acoustic environment in which music students rehearse, whether alone or with an ensemble, must be able to support effectual and meaningful learning, especially during the formative years of their education. Acousticians, architects, audio systems designers, theatrical consultants, building systems engineers, and other members of the design team work together to meet the needs and expectations of the students, music directors, and other stake-holders to provide spaces conducive to learning and practicing music. This section offers a brief overview of the acoustic design process with a specific focus on music education facilities.

Fundamentals of Acoustics

We don't just hear sound; we *feel* it. Music can put a smile on our faces, make the hairs on the back of our necks stand up, and bring tears to our eyes. It moves us, causing visceral feelings such as excitement, happiness, sadness, and surprise. Music activates many regions of the human brain, from the auditory cortex to the prefrontal cortex, involved in higher-order cognition, to the reward processing centers in the basal ganglia [2,3].

How do we hear and feel music? When a musician plays their instrument, the sound is communicated to the listener through *sound waves*. Sound waves are patterns of vibration in a medium (such as air) that travel from one location to another, aptly named the source and receiver locations. The source, e.g. a trumpet or violin, vibrates the air particles around it, generating instances of compression (particles squeezed together) and rarefaction (particles forced apart) in the medium. These air particles in turn vibrate the particles next to them, and so on, propagating the wave to the receivers' ear. Once at the receivers' ear, the wave is shepherded through the ear canal via specially shaped folds and grooves on the ear. The wave excites the air particles inside the ear, which hit the eardrum and cause it to vibrate. The eardrum sends the vibrations on a trip through a series of bones in the ear, continuing through a spiral cavity in the inner ear called the cochlea, ending in the auditory nerve, ultimately producing electrical impulses, which are interpreted by the brain as sound.

For any given vibrating air particle, compression and rarefaction oscillations occur many times over a period. *Frequency*, or the number of cycles of oscillation per second, is used to describe this physical property of the sound wave, and is expressed in *Hertz (Hz)*. The note A4, the A above middle-C, is often used as a general tuning standard to align musicians prior to playing as an ensemble. A4 typically has a frequency of 440 Hz [4] (though, the frequency can vary between 432 Hz and 444 Hz due to local preferences or standards in different parts of the world), which means that the air particles complete a cycle of compression and rarefaction 440 times per second. A listener detects the frequency of a sound wave through the subjective experience of *pitch*. A low-frequency sound wave has a low-sounding pitch, like a cello or double-bass. Conversely, a high-frequency sound wave has a high pitch, like a piccolo.

Acousticians assess sound over a wide range of frequencies because of the large frequency response of the human ear (approx. 20-20,000 Hz). Additionally, many acoustic parameters vary with frequency. It is useful to divide frequencies that are similar to one another into groups, otherwise known as *frequency bands*. These ranges of frequencies are denoted by the frequency in the center of the band. When the center frequencies are separated by factors of 2, such as 125 and 250 Hz, the groups are called *octave bands*. Acousticians typically use octave bands from 31.5-8000 Hz to assess rooms for music, though the range can differ depending on the circumstances. In this book, the acoustical data is shown for octave bands from 125-4000 Hz to remain consistent based on the available data for each facility.

Another important physical property of the sound wave is the extent of the compression and rarefaction. The *sound pressure level (Lp)* with units of *decibels (dB)* describes the amplitude of the sound wave. Sound pressure level decibels are calculated using a logarithmic ratio of air pressure with respect to a specific reference. The reference of 20 micro-Pascals is commonly used, and referred to as the threshold of human hearing (0 dB). A listener perceives sound pressure level as *loudness*. For a listener one meter away, a person calmly breathing has a level of approximately 10 dB, a normal conversation falls in the range of 40-60 dB, while traffic from a busy highway is typically between 80 and 90 dB, and a jet engine is around 130 dB.

Architectural Acoustics

Music also has a profound impact on how we connect with others, allowing us to communicate by transcending language and culture. The connections between performers and between performer and audience is a critical relationship that depends on how sound moves through a space. The field of science that studies the relationship between sound and the space in which it is experienced is called *architectural acoustics*. When designing a space with acoustics in mind, there are three main areas of focus: *interior room acoustics, sound isolation,* and *background noise control*. While the scope of the present book focuses on music education facilities, architectural acoustics design applies to any space where sound can affect the perception and behavior of people, including concert halls, drama theatres, schools, hospitals, worship spaces, residences, and workplaces.

Interior Room Acoustics

The study and design of interior room acoustics concentrates on the characterization and optimization of sound energy within a space. Each space requires an acoustic character that should facilitate the given use of a space. Whether they are intentionally reverberant or dry, spaces should be free of errant, harsh reflections or excessive noise buildup. These room characteristics are impacted by a room's geometry, volume, layout, and surface finishes.

When a sound wave collides with a surface, some of the energy from the wave is reflected back into the room, while the rest of the energy is either transmitted through or absorbed by the surface. The amount of energy that is reflected, absorbed, or transmitted is determined by the material properties of the surface. Surfaces with soft, fuzzy finishes, such as fiberglass, are efficient at absorbing and dissipating the energy of a sound wave because the moving air particles rub against the material, causing friction. On the other hand, hard and dense materials, such as glass or concrete, tend to reflect sound. Thin surfaces, such as wood paneling, tend to allow sound energy to transmit through. It is important to carefully consider the type and placement of each surface to appropriately distribute sound in the room through the use of absorption and reflection.

The identification of the sound source and receiver is crucial for the successful implementation of room geometries and surfaces. However, the source and receiver may differ depending on the type of space analyzed. This book contains two general types of spaces: *performance* and *rehearsal/practice*. The source for a performance hall is the musician on stage, while the receivers are the audience or the other musicians in the ensemble. In practice rooms, the sources are the same, but the intended receivers are often the musicians, conductor, or music director. Both of these conditions pose interesting design problems in how to effectively and pleasurably communicate sound between sources and receivers by supporting reflections that aid communication and reducing reflections that hinder it.

For example, imagine two musicians sitting on opposite sides of a rehearsal room playing together. The sound that travels in a straight line from one performer to the other is known as the *direct sound*. Reflections from the ceiling and sidewalls can be used to support the direct sound so that the musicians can hear each other better. However, if some strong reflections take too long to travel to the other musicians, these reflections may be perceived as *echoes*, which may be distracting and cause the musicians to play out-of-sync. These delayed, high-amplitude reflections may also cause the listeners to perceive the sound source to be located in the wrong place.

The unwanted reflections may be mitigated by either applying absorptive materials, or in the case where reverberation time should be maintained, diffusive treatments. The effectiveness of a material's absorption depends on the thickness, surface area, and mounting conditions. Thicker materials absorb sound more efficiently than thinner materials because the acoustic wave needs to travel through more material, thereby allowing more opportunities for friction to dissipate the energy from the moving air particles. After a sound wave reflects off of a hard surface, it interacts with itself. This *superposition* of the original and reflected waves induces a large particle velocity at a distance of a quarter of the wavelength from the wall and zero particle velocity at the wall. Since it is important to ensure that a large portion of the wave moves in the material to attenuate the sound, a material will be more effective at absorbing sound if it is thicker than a quarter of the wavelength. Conversely, for a given material thickness, sounds with shorter wavelengths (i.e., higher frequencies) are absorbed more easily than sounds with longer wavelengths (i.e., lower frequencies).

There are instances where designers aim to absorb a targeted frequency in the form of tuned absorption. Examples of this are recording rooms and mastering rooms. These small spaces often suffer from low-frequency buildup due to the development of standing waves. Tuned absorption is a useful solution to disrupt the low-frequency standing waves. Where thick, fibrous absorption is broadly effective at middle and high frequencies, tuned absorbers can be designed to target a narrow band, problem frequency, which is often a lower tone. These low-frequency, bass absorbers are generally rectilinear or triangular boxes with a thin, resonating membrane (e.g., 1/8" plywood) fixed to a rigid frame. A 1-2" thick fiberglass wrapped absorptive panel is typically applied to the resonating panel for additional absorption. The thickness of the face is calculated to resonate at the problem frequency. The depth of the box is calculated such that the resonating face, along with the air entrained within, effectively damp the problem frequency. Generally speaking, the volume of the entrained airspace needs to be larger for lower frequencies.

Sound Isolation

Sound isolation, also referred to as acoustic separation, is the practice of preventing unwanted sounds from reaching a critical space. It is a key aspect in fostering focused practice and uninterrupted performances in music education facilities. There are two general practices for achieving sufficient acoustic separation. First, key rooms can be spatially isolated, ideally during the initial planning and programming phases of a project. Keeping acoustically critical spaces well separated from noisy spaces, such as mechanical rooms and bathrooms, greatly reduces the chance of sound leakage into the key room.

If critical spaces cannot be spatially separated due to project constraints, the walls, floors, and ceilings should be designed to reduce the transmission of sound into the acoustically sensitive spaces. Openings, such as doorways or windows, must be carefully detailed because sound travels through the path of least resistance. As an example, the openings for ductwork or electrical equipment should be sealed to prevent airborne sound from entering. The materials of walls, floors, and ceilings should also be carefully considered. For single-panel constructions, one simple implementation to decrease the transmission of sound from noisy rooms is to increase the thickness and/or mass of the construction. However, the effectiveness of this approach is limited because very thick walls would be required to prevent the transmission of low-frequency sound. Therefore, more complex and robust constructions are typically used to maintain necessary acoustic separation. For example, double-panel structures, such as two free-standing gypsum wall board studs or double wythe masonry walls, exploit the resonance of the airspace between the panels, making it more effective at reducing sound, while maintaining smaller total wall thicknesses. By placing sound absorbing materials inside of the airspace, sound transmission can be further reduced.

In addition to airborne sound (i.e., sound that travels through the air), it is also important to prevent structure-borne sound from external sources, particularly vibrational sources such as a passing train, from entering the critical space. Ideally, vibrational noise is attenuated at the source. There are several different ways to implement vibration isolation of machinery, which is often the noise source, such as damping isolation pads and tuned mass-spring support systems. If the source cannot be modified, the transmission of vibrational energy can sometimes be mitigated by isolating the room using mass-spring support systems for the floor or ceiling, or a box-in-box construction. However, these approaches tend to be expensive and cost-prohibitive.

A material or contstruction's ability to reduce transmission of sound can be determined by measuring the *transmission loss (TL)* across the partition. *Sound transmission class (STC)* is a single-number rating of the effectiveness of the reduction of airborne noise based on the TL in each octave band. Higher TL or STC values signify a better ability to reduce sound transmission. The *impact insulation class (IIC)*, similar to STC, is a measure of how well a floor or ceiling reduces structure-borne noise, specifically impulsive impacts on the partition such as footfall.

Background Noise Control

Background noise control generally refers to mitigating noise from building systems, such as the noise emitted by mechanical, electrical, and plumbing (MEP) equipment. Absolute silence is not required in most spaces in a building; however, the background noise level must be low enough so that it does not distract from a performance or practice. Typically, acoustical consultants recommend that the background noise in a music performance space be approximately 15 dB lower than the quietest note being played. This design parameter will help both the performers and audience to hear and enjoy a musician playing a piece at pianissimo. The recommendation for rehearsal rooms and practice rooms is slightly higher: a background noise of 5-10 dB louder than that of a music performance space is acceptable because these spaces are less critical and may even benefit from the masking noise offered by higher background noise levels. Lobbies, offices, lounges, and support spaces, are less noise-sensitive, and can be 10-15 dB louder than rehearsal and practice rooms. However, these spaces should still be controlled to prevent bothersome tones, hums, or buzzes.

The predicted or measured background noise in a room can be expressed in several ways, all of which are based on the octave band sound pressure levels. The sound pressure levels are often represented as a single number rating using one of a variety of methods, including: *Noise Criteria Rating (NC), Room Criteria Rating (RC), Noise Rating (NR), Noise Reduction*, or overall *A-weighted Sound Pressure Level (L_A)*. More information on these rating systems can be found in the Glossary.

Acousticians work closely with the project MEP engineers, and the rest of the design team to assure that concerning noise sources are well controlled. Special consideration should be taken to properly select MEP equipment within the building. It is best to select equipment that output low noise levels. However, if such equipment is not available, must be located above grade, or near acoustically critical spaces, acousticians work closely with the project structural engineer to utilize a combination of vibration isolation elements and structural stiffening to limit vibration from transmitting through the building structure.

Noise from air-handling-units is controlled by re-routing ductwork, conduit, and piping; internal duct liners; duct attenuators; and following velocity guidelines for airflow and liquid flow. The routing of ductwork, conduit, and piping should be carefully reviewed to reduce any unwanted noise and vibration. This review also helps to control potential isolation breaches caused by these services. Internal duct liner is sound absorptive material (typically 1-2 inch thick fiberglass or foam) that lines the internal walls of an air distribution duct. As fan noise travels down the duct, it interacts with the liner and is attenuated. Fan noise is reduced more effectively as the length of internally lined duct increases. Similar to fibrous absorbers within a room, duct liner is generally most effective at reducing mid-to-high frequency noise. Since most problematic fan noise is in the 125-Hz and 250-Hz octave bands (lower frequencies), a passive sound attenuator can replace sections of duct with thick baffles lined with sound absorptive material. Attenuators can either target problem frequencies or act more broadly across a wider range of octave bands when necessary. Finally, the flow rate of air or liquid within ducts or pipes can also cause noise, which will impact the background noise levels in a given space. Slower movement of air and liquid within ducts or pipes yields lower noise levels. In critical spaces, the airflows are designed to be slow and smooth so as not to introduce distracting noise.

Vibration-induced noise can also be controlled by introducing vibration isolation elements at the mounting points of reciprocating or otherwise noise-making equipment. Neoprene isolators provide damping material that dissipates the vibration, while spring isolators can be tuned to match the resonances of the vibrating equipment, thereby attenuating the noise. By choosing the appropriate isolators for the specific equipment, vibration noise can be reduced by up to 90%. Vibration can be further reduced by adding a flexible connection between the vibrating equipment and the associated ductwork, piping, and conduit. This limits the amount of vibration transmitted directly from the vibrating equipment into distribution routes (e.g., ductwork, piping, and conduit).

Additional Resources

As described above, and as shown in many of the music education facilities presented in this book, acoustic design metrics and strategies are often interrelated, such that only a holistic approach will ensure success. The discussion above is only a brief overview of this approach, articulating some of the major aspects involved in designing the acoustics of a space, and is not intended to be a comprehensive design guide. A list of helpful textbooks is given in the References [5,6] for readers interested in a more extensive explanation of the various topics discussed in this overview and other nuances involved in the art and science of architectural acoustics. To learn more about the science of acoustics and to search for educational programs, please visit the Acoustical Society of America website (acousticalsociety.org). Also, more information about acoustical consulting companies, many of whom work in the area of music facility design, can be found through the National Council of Acoustical Consultants (www.ncac.com).

References

[1] Carlson, Matt. "The Joke." *Carnegie Hall*, https://blog.carnegiehall.org/Explore/Articles/2020/04/10/The-Joke. Accessed 25 June 2020.

[2] Poeppel, David, Overath, Tobias, Popper, Arthur N., and Fay, Richard R. (Eds.) *The Human Auditory Cortex*. Springer 2012, New York.

[3] Juslin, Patrick N. and Sloboda, John A. (Eds.). *Music and Emotion: Theory and Research*. Oxford University Press, 2001, New York.

[4] ISO 16:1975. *Acoustics - Standard tuning frequency (Standard musical pitch)*. (Standard No. 3601). Retrieved from https://www.iso.org/standard/3601.html.

[5] Long, Marshall. *Architectural Acoustics*. Elsevier Academic Press, 2006, Massachusetts.

[6] Mehta, Madan, Johnson, Jim, and Rocafort, Jorge. *Architectural Acoustics: Principles and Design*. Prentice-Hall, Inc., 1999, New Jersey.

Reflections
FROM KEY DESIGN TEAM MEMBERS

THE SOUNDSCAPE OF MUSIC REHEARSAL AND EDUCATION FACILITIES
GARY W. SIEBEIN

GARY W. SIEBEIN, FAIA, FASA

Professor Emeritus of Architecture and Acoustics at the University of Florida School of Architecture & Senior Principal Consultant with Siebein Associates, Inc.

Siebein has written five books, 19 book chapters, and over 200 technical papers and presentations on architectural acoustics, environmental sound and soundscape design. He directed a graduate research program where masters and doctoral students from around the world studied and conducted research in building acoustics, environmental sound and soundscape design. He is also the Senior Principal Consultant with Siebein Associates, Inc. (www.siebeinacoustic.com), an acoustical and soundscape design and consulting firm located in Gainesville, Florida that specializes in design for natural and reinforced acoustics in music rehearsal and performance, worship and theatrical spaces; as well as just about every other type of building. He has thoroughly enjoyed working and collaborating with exceptional clients and design teams on over 2,300 exciting projects since the firm's inception in 1981.

Acoustic data which characterize tonal characteristics in sound radiation of musical instruments as well as room acoustic processes are objective facts. Performance practical directions in many cases are only examples of subjective interpretations. They are intended to show possibilities for utilizing acoustic facts for the realization of an artistic tonal perception [1].

An acoustical community is a soundscape where acoustical information plays an important role in the lives of the inhabitants that is characterized by a variety in the kinds and patterns of sounds; a complexity within the sounds and in the levels of information they convey; and a functional balance that operates to constrain variety and complexity [2].

The way a sound functions in a system depends not only on its level and other "objective" characteristics, but in the way it is understood by listeners and the community" [3].

The English word communication is derived from the Latin word communicare which means to make common, to share; to give something intangible to another; to convey in speech, writing, music or signs [4].

The soundscape of music rehearsal is a multi-year, international investigation into the qualitative and quantitative acoustical attributes and design strategies for music rehearsal and education facilities. Academic research and professional acoustical consulting practice in this area can have a significant impact on the design of successful education and practice facilities for music students and professionals. Siebein has defined 5 levels and 7 elements of soundscapes that are pertinent to the design of music education spaces [4,5].

A Brief History

The first acoustical research in this area was conducted by Wallace Clements Sabine in 1902 when he interviewed 5 musicians in small rooms used for music instruction at the New England Conservatory of Music about the qualities of music that they heard in their practice spaces and related these observations to acoustical measurements of reverberation time that he made in the spaces. He incrementally added and took away sound absorbing cushions from the Sanders Theater in the rooms and went through the process again until the musicians determined that the acoustical conditions were satisfactory. He developed a conclusion about reverberation time and practice spaces based on the opinions of the musicians. This was among the earliest studies to link perceptual evaluation of music qualities with the architectural and acoustical features of music practice spaces [6].

Parkin and Humphreys [7], Knudsen and Harris [8] and other early acoustical authors picked up on Sabine's study and provided recommended reverberation times in their books on architectural acoustics in the 1930's through the 1970's. Rein Pirn [9] conducted a very interesting study in rehearsal spaces where he showed that even with an acceptable reverberation time, that loudness would build-up too rapidly in a room that did not have sufficient ceiling height to allow the relative strength of sounds to dissipate somewhat before being reflected back to the musicians. In the 1970's authors such as Beranek [10], Egan [11], Doelle [12] and others provided acoustical guidance on the acoustical design of music education spaces that included shaped wall surfaces for sound diffusion and reverberation time criteria. Wenger [13] published a guide on the design of music education spaces that included reverberation time criteria for vocal and band spaces separately, ceiling height and volume per musician to assist architects in the design of these spaces.

In the 1970's new measurement techniques were being developed to study and design for the individual reflections that constitute the build-up of sound as heard by people in the room. These techniques also improved upon the measurement of the decay of sound or how long the sound persists in the space after the source has stopped that was the basis for Sabine's pioneering work with reverberation time. Siebein [14] showed that acoustical measurements of reflected sounds varied in a 1:10 scale model and in the full-size room of a music education space. Siebein and Cervone [15] showed that these variations in

Five Levels of Soundscape Design

Siebein [4] identified five points in the design process that soundscape theories can inform the design of buildings:

The *inspiration* and underlying philosophy guide the design and help to set the framework within which design occurs. The inspiration for a project occurs before any design actually begins. Musical sounds, the structure and rhythm of music and other metaphors are often used by architects and soundscape designers to contribute underlying ideas to a project at its initial conception.

Planning is the larger scale design ideas that organize the experience of the building along an itinerary in space and time.

A *conceptual structure* is the underlying set of principles and ultimately geometries that form the basis for the intellectual and formal aspects of the project. A conceptual structure is often derived from transformational mapping studies of a site and localized contextual influences such as sounds, weather, climate, social forces, circulation systems, traditions, historical influences, spatial systems, etc. This conceptual structure is ultimately the shaper or giver of sound and the "coloration" it receives from the building.

Tectonics are the elements that form the architectural system that the soundscape occurs within. The tectonic elements are arranged in a conceptual structure derived from the local ecology or interrelationships between elements in a pattern or system that can be mapped in its literal, physical or metaphysical dimensions. The tectonics are those elements that give a unique identity and form to a place.

Details are the local connections among the tectonic elements that support and express the inspiration and the conceptual structure of the project. The details are also the elements that often provide weather protection, connections among structural elements and enclosural systems and elements that allow for sonic and other environmental flows in indoor environments.

Seven Elements of Soundscape Theory [5]

1. An *acoustical community* is a group of people linked by the importance of the communication they exchange such as music students and instructors.

2. Webs of *ecological relationships* structurally connect the members of the community. These members need to hear each other and discern subtle meanings and cues from the sounds of others.

3. *Acoustic calendars* or rhythms occur for each of the sound and activity cycles for each of the participants that usually result in some variations of sounds and activities over a diurnal, monthly, yearly or other cycle. This can be the repetitive cycle of playing difficult passages with different members of the musical ensemble or having the instructor give some preparatory remarks to give focus to the session.

4. *Acoustic rooms* are specific locations with localized sonic events that are uniquely colored or otherwise affected by the surroundings. An acoustic room is a specific location where sonic exchanges occur.

5. An *acoustic itinerary* is a path where the members of each community move, create and listen to sounds during the course of their activities during the rehearsal or class.

6. *Sonic Niches.* Sounds are not made by all members of an ensemble simultaneously, nor are all members playing the same sounds when they do play together. The composer and the music director arrange niches in time, pitch, loudness, location, rhythm that the sounds can inhabit to allow them to be heard alone as in the case of a soloist or in combination with the sounds of others.

7. *Sonic flows* are sounds that occur within, between and among acoustic rooms in the soundscape such as when one section of the ensemble is playing and others are waiting to play. The concept that soundscapes have somewhat permeable edges offers interesting design possibilities.

measurements on a location by location basis were not a statistical variation in sounds, but were related to the different structure of the direct sounds and reflected sounds for each unique source-path-receiver combination at the selected measurement locations. Gade [16] found that musicians on stage wanted to hear each other so that they could play in ensemble. He found that the ability of musicians on a concert stage to hear each other was related to early reflected sound energy from the walls and ceiling of the stage enclosure that arrived within 100 milliseconds to 200 milliseconds of the direct sound using the new impulse response-based measurements.

David Eplee [17] completed a Master of Architecture thesis conducted at the University of Florida that was based on scale model experiments of reflected sounds between musicians on the stage at the Grosser Musikvereinsaal done in collaboration with Kirkegaard. He found that reflected sound paths were necessary for musicians to hear each other because the musicians, the music stands, and their instruments blocked the direct sound path for communication across the stage.

Research Study on the Soundscape of Music Rehearsal

Tsaih [18] built on these studies in her doctoral dissertation that combined qualitative evaluations of music education sessions and rehearsals by students and faculty, documentation of the architectural features of the classrooms, and acoustical measurements to analyze the relationships among the perceived sounds and attributes of the rooms. This linking of the qualitative and quantitative aspects of the acoustical experience of music rehearsal and education is what makes this a soundscape study. She found that:

1. Hearing each other, playing in time, in tune, dynamics and articulation were identified as the qualities of music that music instructors and students listen for during rehearsals and practice.

2. Speech intelligibility was found to be important so students can hear and understand the instructor.

3. Interviews and focus group discussions with college level music faculty; and college and high school music students and questionnaires given to both groups were used to evaluate the acoustical qualities of 3 different music rehearsal rooms.

4. Acoustical measurements were made at near and far locations from musicians in all three rooms to evaluate individual communication paths between the instructor and each group of students as well as between other students located close to and farther away from each group of students.

Current Status of the Work

There is a Norwegian standard [19] on music room acoustics that includes guidance on room volume, ceiling height, absorbent materials, variable acoustic features, background noise and a note about including some sound diffusing surfaces in the rooms. The criteria for the rooms in the Norwegian standard are grouped by the type of music that will be played in the room, the size of the room, loudness and reverberation time among others. ISO currently has a working group looking into developing a standard on this topic. There is also a resurgence of international research as well as significant practice-related work being done on the part of acoustical consultants around the world in this area as evidenced by the number and quality of recent paper sessions at Acoustical Society of America meetings.

Future Directions

The concept of designing a music education and rehearsal space can be accomplished using currently available computer modeling software, acoustical measurement systems, and subjective research methods. Such subjective studies may include focus group studies and questionnaires to members of the acoustical communities involved in the rehearsal and educations spaces, similar to Tsaih's [18]. These techniques open the doors for a new generation of rehearsal and education spaces acoustically tuned so that musicians can hear each other well and play in tune, in time, in dynamic and articulation at both near

Summary of Findings of the Soundscape of Music Rehearsal Study [18]

Acoustical measurements made of direct and reflected sounds at locations of other musicians near the musician and at musicians farther away from the musician were analyzed in conjunction with the responses of the musicians to questionnaires about how well they could hear each other. She found that in rooms where sound reflections arrived from other musicians, hearing each other and the ability for students to play in time, in tune and in dynamics with each other increased.

Analysis of video recordings of college, university and high school music rehearsals showed that about 48% of the time was dedicated to verbal discussions between the instructor and students. This means that the acoustical design of the rehearsal room should accommodate speech communication in addition to musical qualities.

Interviews with 13 music conductors showed that hearing each other, blend of sound, tone quality, intonation, dynamics, rhythm and articulation are the primary qualities of music that instructors listen for during rehearsals and classes.

Questionnaires administered to 206 students in high school college and university music classes said that the qualities of music they listened for during rehearsals were articulation: 95%; intonation: 93%; rhythm: 89%; other player's playing: 86%; dynamics: 73%; and others (tone quality, phrasing and style): 14%.

University level rehearsals focused on more sophisticated playing and listening techniques including articulation, dynamics, rhythm and others while younger students focused on basic issues such as intonation and rhythm.

and far locations from others in the ensemble. The spaces can also be designed to allow the instructor or conductor to hear members of the ensemble and communicate with them visually and verbally when desired.

Music education and rehearsal spaces of the future will be designed not only to have a reverberation time or range of reverberation times suited to the musical genres, instructional levels and musical preferences of the staff, but also to facilitate the individual reflective communication paths so that hearing each other will become a given attribute of music rehearsal spaces. The acoustical design of specific reflected sound paths within relatively narrow time windows in which the reflections should occur will become acoustical elements in the rehearsal and education spaces of the future. Additionally, these designs will be able to provide variable acoustics and electronic enhancement systems when desired. Development of a kit of architectural and acoustical parts for rehearsal rooms is possibly within reach once the qualitative and quantitative aspects of the rooms are understood. Organized acoustical measurements using directional sound sources similar to actual music instruments can be used to identify specific communication paths. Continued efforts at pre- and post-occupancy evaluations of the rooms using focus group discussions, interviews, questionnaires, and other qualitative research methods, as well as acoustical commissioning that addresses these issues, can be more easily included in the design process of these important rooms as knowledge and criteria become more widely understood and standardized.

References

[1] Meyer, Jurgen. *Acoustics and the Performance of Music*, 5th edition, Springer Science and Business Media, 2009, New York.

[2] Schafer, R. Murray. *The Soundscape: Our Sonic Environment and the Tuning of the World*. Destiny Books, 1977, Rochester, Vermont.

[3] Truax, Barry. *Acoustic Communication*. 2nd edition, Ablex Publishing, 2001, Westport, Connecticut.

[4] Siebein, Gary W. "Creating and Designing Soundscapes." Edited by J. Kiang et al., *Soundscapes of European Cities and Landscapes*. COST Office, Soundscape COST-2013, Oxford.

[5] Siebein, Gary W. "An Exploration of the Urban Design Possibilities Offered by Soundscape Theory". *Proceedings of the AESOP-ACSP Joint Congress*, 2013, Dublin.

[6] Sabine, Wallace Clement. *Architectural Acoustics*. Paper No. 2 in *Collected Papers on Acoustics*. Harvard University Press, 1923. Published 1964 by Dover Publications Inc., New York, p. 72-77.

[7] Parkin, Peter H. and Humphreys, Henry Robert. *Acoustics, Noise and Buildings*. Praeger, 1958, New York.

[8] Knudsen, Vern O. and Harris, Cyril M. "Acoustical Designing in Architecture." *Published for the Acoustical Society of America by the American Institute of Physics*. 1950.

[9] Pirn, Rein. "On the Loudness of Music Rooms." *The Journal of the Acoustical Society of America*, vol.

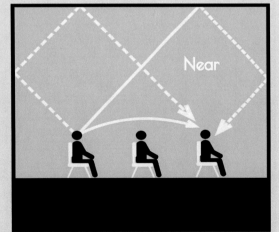

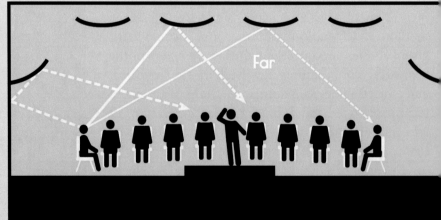

53, no. 301, 1973, p. 301.

[10] Beranek, Leo L. *Noise and Vibration Control.* McGraw-Hill, 1971, New York.

[11] Egan, M. David. *Architectural Acoustics.* McGraw-Hill, 1988, New York.

[12] Doelle, Leslie L. *Environmental Acoustics.* McGraw-Hill, 1972, New York.

[13] Wenger Corporation. *Planning Guide for School Music Facilities.* 2008. www.wengercorp.com/ Construct/docs/Wenger%20Planning%20Guide.pdf.

[14] Siebein, Gary W. *Project Design Phase Analysis Techniques for Predicting the Acoustical Qualities of Buildings.* University of Florida Architecture and Building Research Center, 1986, Gainesville, Florida.

[15] Siebein, Gary W. and Cervone, Richard P. "Listening to Buildings: Experiencing Concepts in Architectural Acoustics." *Education Honors Monograph,* Edited by J. Bilello, American Institute of Architects, 1992, Washington D.C.

[16] Gade, A. C. "Musicians Ideas about Room Acoustic Qualities." *Technical University of Denmark.* Report No. 31, 1981.

[17] Eplee, David F. *An Acoustical Comparison of the Stage Environments of the Vienna Grosser Musikvereinsaal and a Scale Model.* 1989. University of Florida, Master of Architecture Thesis.

[18] Tsaih, Lucky Shin-Jyun. *Soundscape of Music Rehearsal in Band Room.* 2011. University of Florida, Ph.D. Dissertation.

[19] Norwegian Standard NS 8178-2014. *Acoustic Criteria for Rooms and Spaces for Music Rehearsal and Performance.* Standard Norge. ICS 91.120.20.

ARCHITECTURE AND ACOUSTICS IN AN
ERA OF EXPERIMENTATION
CLIFFORD GAYLEY

CLIFFORD GAYLEY, FAIA
Principal, William Rawn Associates

Gayley is a Principal at William Rawn Associates of Boston, known for major public buildings, cultural facilities, and college and university projects. Since joining the firm in 1989, his university work includes projects at University of Virginia, Johns Hopkins, MIT, Berklee College of Music, Haverford, Swarthmore, Bowdoin, Babson, Brown, Northeastern, Vanderbilt, Stanford and Harvard. His performing arts work includes the Rubenstein Arts Center at Duke, the Linde Center for Music and Learning and Ozawa Hall at Tanglewood, the Performing Arts and Humanities Building at UMBC, the Music Center at Strathmore; his civic work includes the Cambridge Public Library and the Boston Public Library, Johnson Transformation. A Fellow of the American Institute of Architects, he has led projects receiving six Honor Awards from the national American Institute of Architects (AIA) and 80 other national and local awards.

Architecture and acoustics are inextricably intertwined, using the same tools (volume, shape, surface) to create places for the exchange of ideas transmitted through sound, movement, and words. This remains true for well-established building types like concert halls, theaters and opera houses. It is also true for new, emerging types that respond to changing client demands.

The Pop-Up

Spaces beyond rehearsal and performance rooms can also be planned as sonic spaces in ways that support impromptu music activity and performances. Lobbies (such as at the Performing Arts and Humanities Building at University of Maryland, Baltimore County) can be reimagined as highly visible venues that take advantage of their typical tall volumes, with careful planning around room proportion, ceiling and wall absorption, and built-in audio-visual infrastructure. At Berklee College of Music, a new type of flexible space - a performance space that during the day serves a 400-seat dining function - provides a unique venue for a pop-up Caf Show concert series: a 5-night-a-week, student-run performance series that celebrates this College's entrepreneurial focus on students developing their artistic brands. Such frequent and radical flexibility requires careful thinking around ease in turnover, including: acoustic infrastructure of reflective and absorptive surfaces and tall volume, plug-and-play A/V system around the stage, lighting grid with well-developed presets, low stage (2 steps up with cabling integrated to floor boxes), upper level dining balcony for added audience capacity and for late dining to overlap during changeover to performance mode.

The Cauldron

Universities seeking to foster interdisciplinarity and creativity are repositioning the arts and developing new types of facilities that encourage transdisciplinary activity, within the Arts and across all disciplines. A key challenge with these facilities is creating real places of collision where researchers, artists, and collaborators in one space are aware of the work of other teams; mere co-location is not enough. At Duke's Rubenstein Arts Center, a non-departmental, provost-led facility, twelve studio labs (six for sonic and movement activities, six for visual art activities) meet that challenge with a system of double barn doors and large acoustic windows that allow each studio to regulate how and when it opens to and helps activate the shared interior hub or lobby. Studio labs can be physically open (with both barn doors retracted on either side of the window), visually open (with both barn doors deployed in front of the opening) and closed off for privacy when needed (with the inner door in front of the window and the outer door in front of the opening). Visible through the large acoustic glass windows, activities within each studio lab animate the life of the building and strengthen community across teams. With both barn doors physically open to create a large portal, studio labs can transform into an extension of the shared lobby welcoming visitors or other teams informally for some shared activity. Creating this flexibility of aperture required careful architectural and acoustic coordination.

The Mashup

Learning institutions are experimenting with ways to bring a building's most active spaces together in one place in exciting ways, requiring careful acoustic attention. At the Boston Public Library, a new type of space – the Big Urban Room – deploys a mash-up of uses to engage the life of the library with a busy retail corridor. An absorptive ceiling prevents sound created by one activity from disturbing other adjacent activities. Included in this continuous and open Big Urban Room is a café and public radio studio (WGBH) at a prime corner overlooking Boylston Street. Over time, the café and studio – the most dynamic parts of this experiment – have thrived, becoming a theater where people come to watch radio live, whether NPR or bluegrass concert, right in the heart of the library.

CAF Show, Berklee College of Music | Robert Benson Photography

Big Urban Room, Boston Public Library | Robert Benson Photography

Duke University, Rubenstein Arts Center | Robert Benson Photography

The Inversion

As campuses expand from their cores to define new precincts around forward-looking goals, rethinking the multi-use hall (for music, theater, dance and spoken word) can be an important ingredient to new precincts. At Harvard Business School, Klarman Hall defines a new type of convening hall at the heart of Harvard's evolving Allston Campus. It enables uniquely intimate conversations between presenters and participants (audience members are elevated to participants in this space) and among participants around pressing issues and problems facing society today. To do so, Klarman Hall experiments in inversion – inverting the acoustic formula of the concert hall (reverberant hall damped by variable acoustics as needed) to prioritize the spoken word (in an acoustically dry hall with reverberance added electro-acoustically as needed). An adjustable and transparent acoustic ceiling over the stage reshapes natural room acoustics around three occupancy parameters that maximize utilization: 300 seats, 600 seats, 1000 seats. For 300 and 600 seat use, a voice lift system is deployed from the ceiling that allows participants to speak audibly without waiting for a hand-held mic to be passed to them. An electro-acoustic system tunes the reverberance of the hall for music events ranging from soloist to jazz ensemble to full orchestra. A 70-foot by 18-foot high-definition video screen serves convening uses, music uses, and theater uses (as a virtual proscenium arch). Such experiments stretch the collaboration between architects, acousticians, A/V consultants and theater consultants in exciting ways.

The Immersion

Learning spaces, whether for incubation, rehearsal or performance, can create immersive experiences tied both to place (connecting to daylight and views) and to creative process (revealing how a work is made). The Tanglewood Learning Institute's Linde Center for Music and Learning experiments with immersing its programming in Tanglewood's memorable landscape as it curates multi-day, cross-cultural activities bringing artists and audiences together in new ways, while eliminating any sense of back-stage. Three studios (small, medium, and large) open musicians and audience to views and balanced daylight on multiple sides. Studio E, the largest, features the landscape as backdrop to performance, with a continuous glass curtain wall as the upstage wall. Oversized barn doors directly connect the stage and the lobby introducing an unexpected and palpable informality – with audience entering and exiting the room from the stage (same as performers) and with the lobby morphing into the off-stage area during performance. A shared Café brings audience and performers together around long farm tables. In these ways, learning and artistic exploration can be tied to sense of place and a broader community.

For architects and acousticians, experimentation that follows clients' mission-driven goals can open new ways of thinking about how we come together around ideas, music, and the arts in both place and time.

Opposite page: Adjustable Acoustic Ceiling, Klarman Hall | Robert Benson Photography

Above: Studio E within Tanglewood Landscape | Robert Benson Photography

L. M. Ronsse et al. (eds.), *Rooms for the Learned Musician*

Audio/Video System Designers as Interpreters

TIM PEREZ

TIM PEREZ, CTS-D
Senior Consultant, Threshold Acoustics

Perez is a Senior Consultant specializing in Audio/Video systems at Threshold Acoustics in Chicago, Illinois. With a background in electrical engineering and architectural acoustics, he seeks to apply technological solutions in a context that considers our cognitive framework and human perception. He resides in Garfield Park, Chicago, with his wife and three cats.

Audio/video system designers are responsible not only for producing design documents, but also for advocating for the occupants and users of a space. While respecting and supporting the architect's vision is the prime concern, a successful renovation or new construction project must simultaneously respond to the users' needs and stakeholders' vision. We want to understand and hear from those who will occupy the spaces we help design; however, the equipment and technical details of AV systems are familiar and interesting to few. What is most important to the future occupants of an educational space is that it lets them accomplish their teaching goals and supports students' education and growth. To that end it becomes the AV Consultant's role to take what may be lay descriptions of technical needs and, through a process of interpretation, translate these into integrated systems of equipment and infrastructure, designed in concert with the architectural context while supporting the underlying goals of the space and its users.

The process for AV designers begins with a review of the available information: vision statements, budget estimates, program requirements, photographs of existing spaces, et al. With an understanding of the owner's intent and the architectural basis for the project, we can begin to investigate the details related to AV systems. Educational spaces are often parts of a larger institution with existing standards for building projects, information technology, and sometimes AV systems that are critical to understand, so we ideally look for opportunities to communicate with these parties throughout the planning and design phases. With that underpinning, we then seek to engage directly with the users of the space, hoping to understand their experiences with existing systems and wishes for new ones. In a sort of interview process, we try to unpack the ways that current facilities serve or hinder faculty members' and students' work while we further define critical design parameters like ensemble types and sizes, the technical experience level of staff and students, or any particular emphases the institution might have (a competition-winning show choir, for example).

At this point discussions with users often launch into an alphabet soup of technical jargon, when some consultants are inclined to discuss details related to equipment and systems that goes over the heads of many daily users. We instead focus the conversation on the user experience, encouraging a vocabulary of activities and functionality and eschewing equipment manufacturers, models, and specifications whenever possible. This approach sparks more interest among a wider audience of faculty and users and allows people to speak freely in service of defining the AV system goals of the project (along with other systems, whose designers can equally gain insight through this process). We can ultimately gain a deeper understanding of what the space needs to do for the people who will be using it.

We must be nimble and responsive to the many moving parts that interrelate throughout any building project as we proceed through the architectural design phases. On any building project, as each discipline is defined and detailed and the architect's vision is translated into a buildable design, there is an ongoing process of compromise and interplay. It is our duty to help realize these architectural goals while ensuring the systems we design are functional for the users who will eventually occupy these spaces. By internalizing the underlying functional goals for these systems, we as designers are free to explore a wider array of creative solutions that may have been precluded by a prescriptive discussion of equipment and infrastructure during the planning phases. We can work within the architectural context and respond to the needs of our design team partners, translating functional goals into fully defined and documented systems and equipment. Approaching this process with an openness to compromise results in more fruitful collaboration and a better outcome for all.

An important aspect of this relationship with the users and design team is communication. I often tell users that I do not want to "do a design *to* them," but to design *for* them and provide a solution that is responsive to their way of working, teaching, and learning. We can engage in an active and iterative process with users and the design team that makes clear to both parties the impacts of the other's needs, demystifying for both that which is so often obscured in jargon and technical language. We are essentially in a position of customer service and managing the customer's expectations is a key tenet. Our customer, the architect, stays informed of what the users' desires mean for AV systems; our other customer, the user, keeps abreast of the developing architecture and how our proposed solutions can serve their needs. When the users finally have a chance to step into a completed space and experience their new AV systems for the first time, their knowing what to expect can make the move-in process smooth and seamless.

SUPPORTING THE STUDENT IN THEATRE DESIGN
SCOTT CROSSFIELD

SCOTT CROSSFIELD
Director of Design, Americas, Theatre Projects

With a career spanning 25 years, Crossfield has steadily become a worldwide authority on theatre design. As Director of Design, Americas, for Theatre Projects, the world's leading theatre consultancy, Scott helps keep the firm at the forefront of theatrical design and technical innovation. Prior to joining Theatre Projects, Scott was a partner of Davis Crossfield Associates in New York City. He is a member of the American Society of Theatre Consultants, and speaks frequently at conferences on Theatre Design and Theatre History

n almost 25 years as a Theatre Consultant, I have had the honor to contribute to the design of over 500 performance spaces. Like most theatre designers, a sizable portion of my work has been spaces for music education. Along the way, I have learned a few key lessons that form the foundation of my design process. Some of these lessons were learned through many years working in the theatre myself. Some come from working with great acousticians, architects, and colleagues. Others stem from my time talking to educators and students, working with university and school leadership, and time spent in the theatre seeing the principles in action. By following these basic principles, it is possible to create performance spaces that nurture and support the student musician's growth.

When I start shaping any performance space, I start with size and scale. What is the smallest room that meets the project's needs? A common mistake is to size a performance space for the largest annual event, the proverbial "building the church parking lot for Easter Sunday" [1]. Concert and recital halls are musical instruments and performers interact with the space. They hear themselves, they hear each other, and they learn to play in unison as an ensemble. This is critical to their training and growth. When a room is too large, the energy flows into the deep recesses of the space and does not return to the performer. The room is lifeless, unresponsive, and does not speak to the artist. This lack of feedback denies the student musician opportunities to grow and learn.

Imagine a young student learning to play an instrument or to control their voice. They must be able to reach out into the house and grab the audience by their heartstrings. This level of control is difficult to achieve early in your career when you have not yet developed the presence or technique to fill a room. Performing in smaller rooms helps, as it reduces the need for students to strain their voice or instrument to be heard. The room should support the young artist, giving them every possible advantage to fill the space with sound and take the audience on a journey of discovery with them. Always scale the room to the student, never force the student to fill the room. Placing the student first is what makes for a successful educational performance space, not the other way around. Not only are smaller performance spaces better for supporting the artist, they are also less costly to build, operate and maintain. It's a win-win-win.

Once the size and scale are correct, I focus on intimacy. Intimacy is created by the room shaping: getting the form and proportion correct. The room shape must support two critical relationships: The actor-audience relationship and the audience-audience relationship. Unlike cinema or television, both of which have the power to zoom in on the smallest detail of the performer's face, the scale of live performance is the human body: the solo performer. So, when I am designing spaces for music education, I work with the team to scale the room down to one singular person. I often imagine a young student, standing center stage...feeling exposed, performing in front of an audience for the first time. What can I do to support them?

I can support the performer and strengthen their relationship with the audience by wrapping the audience in a tight embrace, shaping the room in such a way that I'm emotionally extending the audience chamber onto the stage, creating a sense that "we're all in one room — we're in this together." It's about community — that bond that ties us all together. To reinforce this, I stack and wrap the audience, bringing them physically closer to the performer to help make an emotional connection. I then tighten the stacked balconies so they are as close as I can make them without compromising the acoustics and sightlines. Some of the audience is above eye line, some is below, and this interplay provides the artist with the opportunity to connect with audiences in three-dimensional space. They will find this skill immeasurably helpful later in their careers if they are lucky enough to tour through many different venue types.

The wonderful thing about enveloping the performer with a vertically stacked and wrapped audience is that the room will feel good without a full house. I believe that when done right, you can "cut the house in half" — meaning it will still feel great with only half an audience. The intimate inner room shaping,

and smiling faces along the balcony fronts, support the student while hiding the empty seats behind. The vertical stacking provides opportunities to break the audience into smaller seating sections so empty seats are less noticeable. The ability to seat the audience in the orchestra level, turn off the balcony lights, or even close the balconies is a finishing trick. This is in stark contrast to a large fan-shaped auditorium, where a sea of empty seats is immediately noticeable to a performer, and there is little you can do about it.

The final critical ingredient to create a space for music is silence. One of my favorite quotes on this topic is by Leopold Stokowski: "A painter paints his pictures on canvas. But musicians paint their pictures on silence. We provide the music, and you provide the silence." We must make quiet spaces that musicians can "paint on." While it is an acoustician that leads the charge, we help with room shaping, proper building planning, and room adjacencies. When done right — magic happens.

Musicians practice tirelessly to perfect their craft. Theatre designers and acousticians must do the same. Our work as designers supports their work as artists, so we must get it right, every time. By building intimate and supportive acoustic environments, we support artists and help connect them to their audiences. I believe it is our obligation to support future generations of musicians and educators by creating for them the best possible performance spaces that we can.

References

[1] Caller. "You Don't Build a Church for Easter Sunday: A Good Lesson for Contact Center Capacity Planning." *Contact Center 411,* 24 April 2014, http://contactcenter411.com/you-dont-build-a-church-for-easter-sunday-a-good-lesson-for-contact-center-capacity-planning/. Accessed 25 June 2020.

LIST OF CONTRIBUTED VENUES

Primary & Secondary Music Education Facilities

Academy of the Holy Names
Tampa, FL I USA

Berkeley Preparatory School
Gries Center for the Arts and Sciences
Tampa, FL I USA

Deerfield Academy
The Hess Center for the Arts
Deerfield, MA I USA

Dr. Phillips High School
Orlando, FL I USA

Edgewater High School
Orlando, FL I USA

New Trier Township High School
Winnetka, IL I USA

Phillips Exeter Academy
Forrestal-Bowld Music Center
Exeter, NH I USA

Rainey-McCullers School of the Arts
Columbus, GA I USA

RI Philharmonic Music School
The Carter Center
East Providence, RI I USA

Trinity Preparatory School
Winter Park, FL I USA

Union Public Schools
Fine Arts Building
Tulsa, OK I USA

The Winsor School
The Lubin-O'Donnell Center for the Performing
Arts, Athletics, and Wellness
Boston, MA I USA

ZUMIX
East Boston, MA I USA

Higher Music Education Facilities: Completed 2000-2014

Bethune-Cookman University
Julia E. Robinson Memorial Music Hall
Daytona Beach, FL I USA

Casper College
Music Building
Casper, WY I USA

Earlham College
Center for the Visual and Performing Arts
Richmond, IN I USA

Hillsborough Community College
Performing Arts Building
Tampa, FL I USA

Mississippi University for Women
Poindexter Hall
Columbus, MS I USA

North Central College
Fine Arts Center
Naperville, IL I USA

Peninsula College
Maier Hall
Port Angeles, WA I USA

Reed College
Performing Arts Building
Portland, OR I USA

Rensselaer Polytechnic Institute
EMPAC
Troy, NY I USA

Scripps College
Performing Arts Center
Claremont, CA | USA

University of Florida
Music Building
Gainesville, FL | USA

University of Florida
Steinbrenner Band Hall
Gainesville, FL | USA

The University of Kansas
Murphy Hall: Choral Rehearsal Room
Lawrence, KS | USA

University of Maryland
The Clarice Smith Performing Arts Center
College Park, MD | USA

University of Music Karlsruhe
Campus One
Karlsruhe | Germany

The University of Tennessee Knoxville
Natalie L. Haslam Music Center
Knoxville, TN | USA

Wenatchee Valley College
Music and Art Center
Wenatchee, WA | USA

Western Connecticut State University
Visual and Performing Arts Center
Danbury, CT | USA

Westminster Choir College
Marion Buckelew Cullen Center
Princeton, NJ | USA

Xavier University
Edgecliff Hall
Cincinnati, OH | USA

Higher Music Education Facilities: Completed 2015-2020

Carleton College
Weitz Center for Creativity
Northfield, MN | USA

DePaul University
Holtschneider Performance Center
Chicago, IL | USA

Eastern Connecticut State University
The Fine Arts Instructional Center
Willimantic, CT | USA

Hope College
Jack H. Miller Center for Musical Arts
Holland, MI | USA

Missouri State University
Ellis Hall
Springfield, MO | USA

Nazareth College
Jane and Laurence Glazer Music Performance Center
Rochester, NY | USA

Northwestern University
Ryan Center for the Musical Arts
Evanston, IL | USA

Olympic College
College Instruction Center
Bremerton, WA | USA

The Pennsylvania State University
Music Building I
University Park, PA | USA

South Dakota State University
The Oscar Larson Performing Arts Center
Brookings, SD | USA

Southeastern University
College of Arts & Media
Lakeland, FL | USA

University of CO at Colorado Springs
Ent Center for the Arts
Colorado Springs, CO | USA

The University of Iowa
Voxman Music Building
Iowa City, IA | USA

The University of Kansas
Murphy Hall: Swarthout Recital Hall
Lawrence, KS | USA

University of Massachusetts Boston
University Hall
Boston, MA | USA

University of Nebraska Omaha
Janet A. and Willis S. Strauss Performing Arts
Center
Omaha, NE | USA

The University of Oregon
Berwick Hall
Eugene, OR | USA

Whitworth University
Cowles Music Center
Spokane, WA | USA

Yale School of Music
Adams Center for Musical Arts
New Haven, CT | USA

Music Conservatories; Music Rehearsal & Community Centers

Boston Conservatory at Berklee
Richard Ortner Studio Building
Boston, MA | USA

Curtis Institute of Music
Lenfest Hall
Philadelphia, PA | USA

Curtis Institute of Music
Wyncote Organ Studio
Philadelphia, PA | USA

Indian Hill
Music Center
Littleton, MA | USA

Manhattan School of Music
New York, NY | USA

Mount Royal University
Bella Concert Hall
Calgary, Alberta | Canada

Music Academy of the West
Hind Hall
Santa Barbara, CA | USA

Musikinsel Rheinau
Rheinau | Switzerland

San Francisco Conservatory of Music
San Francisco, CA | USA

University of Cincinnati
College Conservatory of Music
Cincinnati, OH | USA

University of Melbourne
The Ian Potter Southbank Centre
Southbank, Victoria | Australia

Wheaton College
The Armerding Center for Music and the Arts
Wheaton, IL | USA

Yong Siew Toh Conservatory of Music
Yong Siew Toh Conservatory of Music Building
Singapore

MUSIC EDUCATION FACILITIES

PRIMARY & SECONDARY EDUCATION FACILITIES

As one of Tampa's leading preparatory schools with a co-educational elementary and middle school and an all-girls high school, Academy of the Holy Names offers a well-rounded educational experience, but needed a performing arts facility to house their arts programs, to engage professional-level instructors and guest performers, and to successfully compete with other private schools in the area. Opened in 2017, the Bailey Family Center for the Arts is a two-story, 189,000 sq. ft. multi-purpose performing arts facility that includes an intimate 350-seat theatre with a balcony, learning spaces for instrumental music, choral music, theatre, dance, and the visual arts that are used for the co-educational elementary and middle school students and the all-girls high school students.

Auditorium | Harvard Jolly Architecture

Rear wall | Siebein Associates, Inc.

The Theatre includes parterre and balcony seating in a small room that makes an incredibly intimate performance space. Space shaping design throughout the Theatre enhances the natural acoustics of the hall for band, orchestra and choral concerts. Theatrical productions, "Broadway" type shows, lectures, and cinematic presentations are made possible through the design and installation of the state-of-the-art audio/visual system. Concealed acoustic drapes can be deployed along the rear wall to allow for variable acoustics, making the Theatre equally suitable for the performances of all of the arts programs, faculty and guest artists at Academy of the Holy Names. Located in the center of the facility and nestled between the arts and performing arts classrooms, the Theatre is acoustically separated from the surrounding spaces with expansion joints, a perimeter corridor that buffers the theater from surrounding spaces, and sound lock vestibules.

The acoustical and architectural design of the room consists of a series of convex curves that are specifically shaped to envelope and immerse the audience members with diffuse sound reflections to provide uniform loudness at all seats of clear, intimate and enveloping sounds. This integration of shaped ceiling and wall surfaces required a diligent, collaborative effort with the architect, interior designer and sound contractor throughout the design and construction phases.

The 'acoustical throat' consisting of the gentle curvature that frames the proscenium opening on the sides and overhead provides early reflections that are propagated throughout the entire audience area. The series of curved ceiling panels and the curved side walls progressively bring the softened sound reflections to the ears of the audience so that all seats including those on the balcony and the under balcony are enveloped with sound that arrives from overhead and from the sides. Even the balcony fronts, undersides of the balcony, and the rear walls are shaped to evenly propagate diffuse sounds to provide an intimate and immersive listening experience.

The acoustical design of each music, theatre and dance classroom is tailored specifically for the individual arts discipline. Tuned to provide the appropriate volume and reverberation for music, theatre, and dance, finishes were selected in accordance with a vibrant color pallet to engage the students and enhance the learning environment. Strategic design of the sound isolation assemblies and support space placement offered an economical solution to provide the necessary noise reduction between the adjacent classrooms and practice rooms.

A palette of prefabricated diffusing and absorbing panels of varying depth, high performance acoustical ceiling tiles and hard surface ceiling tiles was used in the Band Room, Choral Room and Dance Rooms. The sound field in each room was intelligently sculpted to accommodate both clarity of speech and music with the reverberance desired for each room within the large room volumes. The Band Room is acoustically designed to accommodate both the subtle sounds of flutes and violins and the robust sounds of brass and percussion instruments through controlled installation of both sound absorbing and sound diffusing panels on the ceiling and walls for communication among the students as well as between the students and the instructor while simultaneously controlling the build-up of loudness in the room. The Choral Room is designed for reverberance that enhances the presence of the choral ensemble as a whole but also provides individual identification of each voice for the choir director through strategic placement of sound diffusing panels on the ceiling and walls so that the instructor can hear each student and the students can hear each other. The Dance Classrooms in which both instrumental and recorded music is played have variable acoustic drapes that can be tuned to control amplified sound playback as well as enhance natural acoustic instruments.

With rooftop air handling units located above the classroom areas and adjacent to the exterior wall of the Auditorium and stage house, careful attention was given to the acoustical design of the HVAC system and the exterior sound isolation assemblies for the Bailey Family Center for the Arts. The duct paths serving the acoustically sensitive spaces, including the Theatre, music classrooms and theatre classrooms were extended to the extent possible. All variable volume terminal units were located in less sensitive spaces such as storage spaces and corridors. The background noise level data shown were measured in the unoccupied space with the HVAC system on. The reverberation time data shown were measured in the unoccupied space.

In addition to the architect and acoustical consultant, the design team included: TSG Design Solutions Inc. (theatre consultant), Carastro & Associates (mechanical engineer), and The Beck Group (general contractor).

BALCONY VIEW | SIEBEIN ASSOCIATES, INC.

BAND ROOM | HARVARD JOLLY ARCHITECTURE

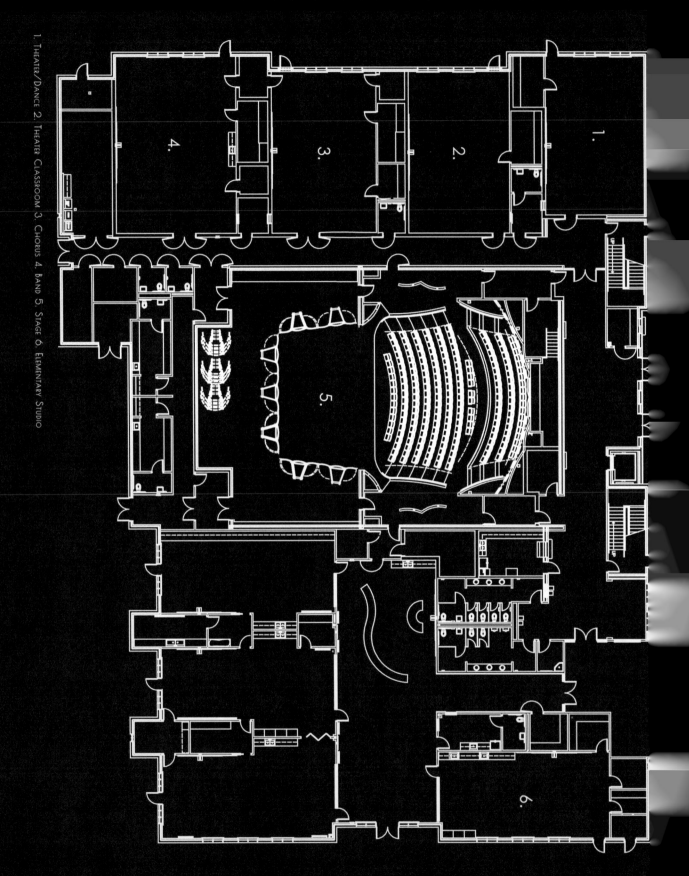

1. Theater/Dance 2. Theater Classroom 3. Chorus 4. Band 5. Stage 6. Elementary Studio

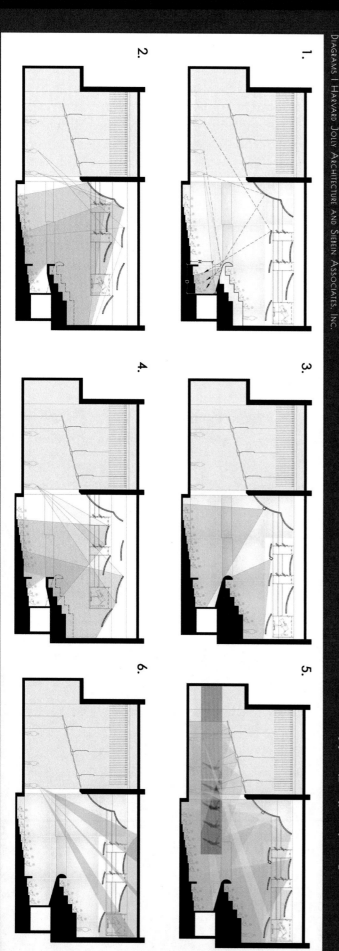

Diagrams | Harvard Jolly Architecture and Siebein Associates, Inc.

1 Reflections—Balcony 2 Reflections—Acoustical Throat 3 Loudspeaker Locations 4 Reflections—Reflectors Over Audience 5 Spaciousness Envelopment 6 Coupled Space Above Suspended Canopy

ACOUSTICAL CONSULTANT:
SIEBEIN ASSOCIATES, INC.

ARCHITECT:
HARVARD JOLLY ARCHITECTURE

COMPLETION DATE:
2017

LOCATION:
TAMPA, FL I USA

CONSTRUCTION TYPE:
NEW CONSTRUCTION

CONSTRUCTION/RENOVATION COST:
$12,500,000

FEATURED SPACE DATA:

ROOM VOLUME:
182,922 ft3

FLOOR PLAN AREA:
6,100 ft2

SEATING CAPACITY:
350

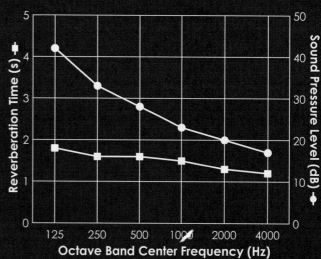

Berkeley Preparatory School is one of the finest private college preparatory schools in Florida offering its students world-class educational opportunities augmented by state-of-the-art technology and a growing fine arts program. Opened for the start of the fall semester of 2015, the Gabos Family Recital Hall is part of the Gries Center for the Arts and Sciences as one of the newest additions to the Berkeley Preparatory School campus. Berkeley's Visual and Performing Arts Program offers facilities carefully designed to enrich all educational and performance aspects of art, drama and music.

The Gabos Family Recital Hall was designed to accommodate a variety of orchestra, band, solo and small ensemble instrumental and vocal music recitals, as well as other activities such as lectures and presentations. Accommodating such a variety of uses requires special acoustical design to allow the space to be configured according to the size and type of ensemble or performance. The hall is home to one of the school's Steinway Model B grand pianos for use in performances and rehearsals. Both natural and amplified acoustical rehearsals and performances occur in the room.

The architect made a bold move to include eight large floor-to-ceiling windows at the rear of the performance platform that had to be accounted for in the acoustical design of the room.

The basic acoustical design concept for the room was to provide enough volume, to shape the surfaces of the room, and select materials with appropriate sonic properties so that the room worked as an

"instrument" to enhance the sounds played within it at all seats in the audience seating area. The audience sits in the same volume as the performers allowing a very intimate performer/audience relationship.

The shaping of the side and rear walls that are exposed when the drapes are retracted, along with the overhead reflecting panels help to provide reinforcement for natural acoustical performances as well as allowing the performers to hear each other when they are playing as an ensemble.

The side wall diffusing surfaces were brought in at the rear of the room to reduce the width and provide diffuse lateral sound reflections to a greater area of the audience seating area to create a sense of envelopment and to enhance the acoustic and visual intimacy of performances.

Continental seating was used to allow more seats to be in the center of the room providing better viewing angles and preferred acoustical qualities.

The sound diffusing wall and ceiling shaping helps to provide strong early sound reflections and a smooth decay of sounds to increase clarity of natural acoustic performances and with multiple early reflections increasing the apparent loudness and providing more uniform sound levels throughout the room with a mid-frequency reverberation time that can be varied between 1.2 seconds to 0.8 seconds when drapes are deployed. The reverberation time data shown were measured in the unoccupied space.

Acoustical drapes were included as an essential component of the acoustical design of the room that can be retracted behind the sound diffusing panels to allow the occupants to vary the acoustical response of the room when desired.

The inclusion of eight large windows to the outside was part of the design that provides a backdrop for performances and allows a connection to the natural environment that surrounds the building for the students. This required laminated insulated windows to reduce noise to the background sound levels of NC 25-30. Storage for the adjacent band room is located under the risers of the recital hall to maximize the use of space. Sound isolation between the recital hall and adjacent music education spaces was designed to allow simultaneous use of the spaces. Acoustical design of the HVAC system included selection of a relatively quiet AHU, inclusion of silencers in the supply and return air ducts, and maintaining relatively low air velocities in the ducts to reach NC levels of 20-25.

Opposite page: Front View | Siebein Associates, Inc.

PERSPECTIVE VIEW | SIEBEIN ASSOCIATES, INC.

CEILING PANELS | SIEBEIN ASSOCIATES, INC.

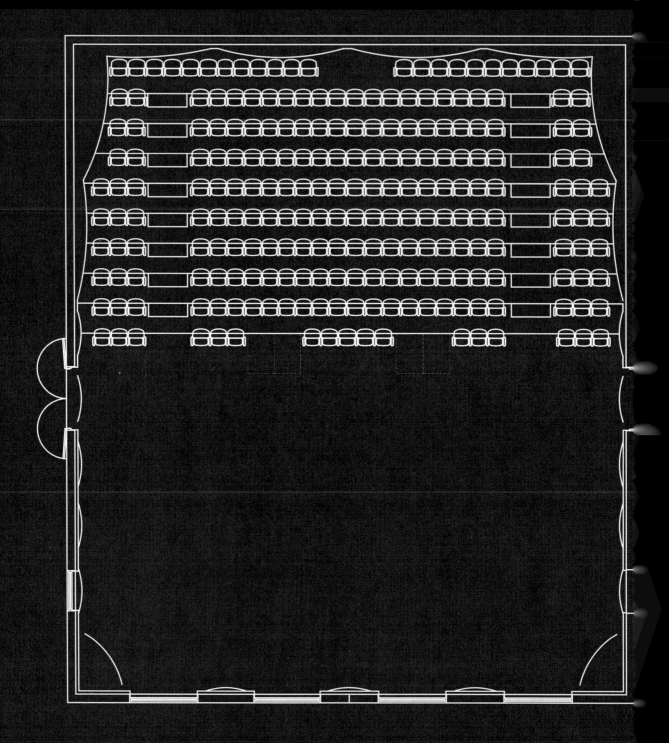

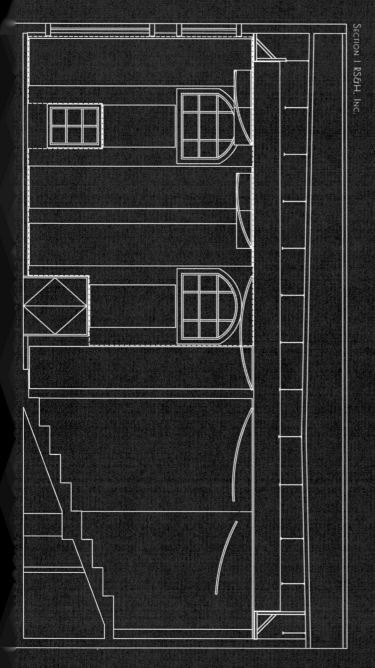

Section | RS&H, Inc.

ACOUSTICAL CONSULTANT:
SIEBEIN ASSOCIATES, INC.

ARCHITECT:
RS&H, INC.

COMPLETION DATE:
2015

LOCATION:
TAMPA, FL | USA

CONSTRUCTION TYPE:
NEW CONSTRUCTION

FEATURED SPACE DATA:

ROOM VOLUME:
72,280 FT3

FLOOR PLAN AREA:
3,555 FT2

SEATING CAPACITY:
230

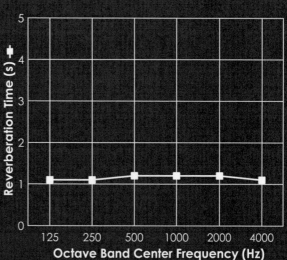

The 85,000 square-foot addition to Deerfield Academy's Hess Center for the Arts was designed to support the demands of Deerfield's growing music, theatre, dance, and arts programs. The addition included a 160-seat concert hall, five new arts studios, and a new art gallery. The existing auditorium received major renovations, including increasing the seating from 700 to 850 seats while also upgrading its lobby, stage technology, lighting, and acoustics. Furthermore, the renovations provided a new music recording studio, numerous new music practice rooms, a second dance studio, and new infrastructure for a black box theater.

Specifically designed for classical music performance and pedagogy, the new and striking 160-seat Elizabeth Wachsman Concert Hall is both intimate and warmly resonant, and is ideal for choral, chamber, and small orchestral music. Diligent sound isolation and an extraordinarily quiet mechanical system ensure that noise does not compete with the music.

The acoustical design of Wachsman Hall relies on its narrow footprint (42 feet at the widest point), generous height (35 feet over the stage), and steep seating slope to provide clarity and intimacy together with intense warmth and reverberation, and remarkable dynamic range. Expressed pilasters and other wall protrusions, together with an inverted coffered ceiling scheme, provide helpful diffusion of sound while maintaining the strong specular reflections so important to clarity and impact. An operable curtain covers the upstage wall up to ten feet when extended, allowing on-demand control of loudness onstage and increased

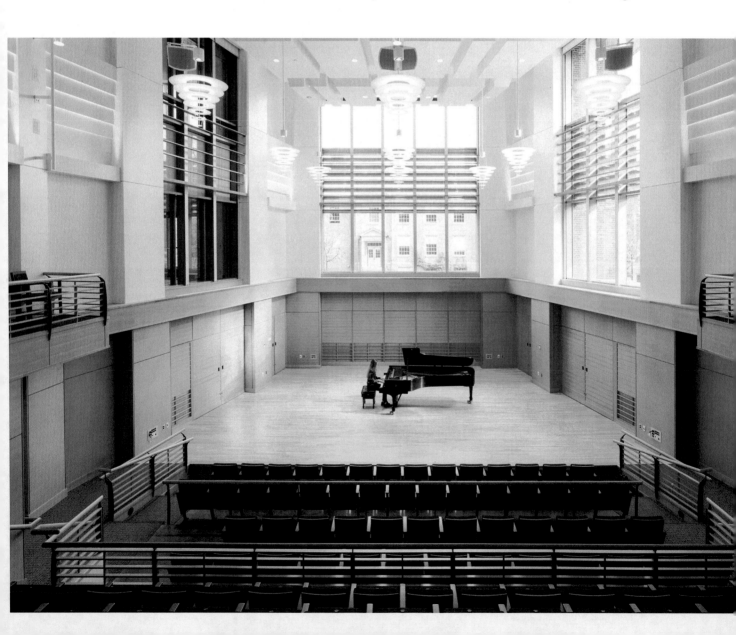

CONCERT HALL – SEATING | EMILY O'BRIEN

BAND ROOM | EMILY O'BRIEN

AUDITORIUM | EMILY O'BRIEN

clarity for larger and louder ensembles.

Large windows flank the stage on three sides at the upper two levels. One of these windows faces an interior corridor, opposite the band rehearsal room. At this location, a double window assembly was used with a deep airspace between thick glass lites.

The renovated auditorium increases seating capacity and improves intimacy by adding side balcony seating at two levels. Stair access is provided at various points between levels and to the rear balcony. The existing ceiling, which was too low, was demolished to reveal structure, facilitate catwalk access, and increase acoustical volume. Mechanical equipment was located in a rooftop penthouse away from the auditorium, improving HVAC function while reducing noise.

The project also includes several new and renovated classrooms, music recording facilities, instrument rehearsal rooms, and art and architectural studio classrooms.

The audiovisual systems in the auditorium and the concert hall each include large, video projection display systems compatible with today's digital video standards and sound reinforcement systems tailored specifically to each space to optimize speech intelligibility and audio quality. The audiovisual design also consists of video displays and audio playback systems for each classroom, wired and wireless backstage communications, and video and audio monitoring of the black box theatre's performance area in the lobby and green room.

In addition to the architect and acoustical consultant, the design team included: Martin Vinik Planning for the Arts (theatre consultant), Acentech Incorporated (audio systems designer, video systems designer), LAM Partners (lighting designer), TMP Consulting Engineers (mechanical engineer), Daniel O'Connell's Sons (general contractor), and Boston Building Consultants (structural engineer). The reverberation time data were measured in the unoccupied space. The background noise level data were measured in the unoccupied space with the HVAC system on.

Opposite page: Concert Hall - Stage | Emily O'Brien

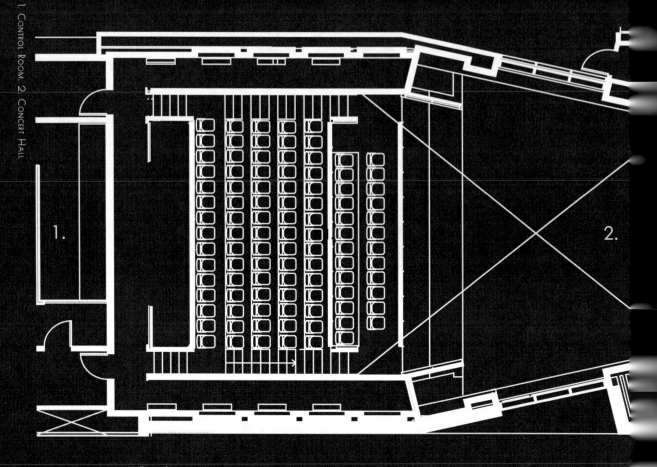

1.

2.

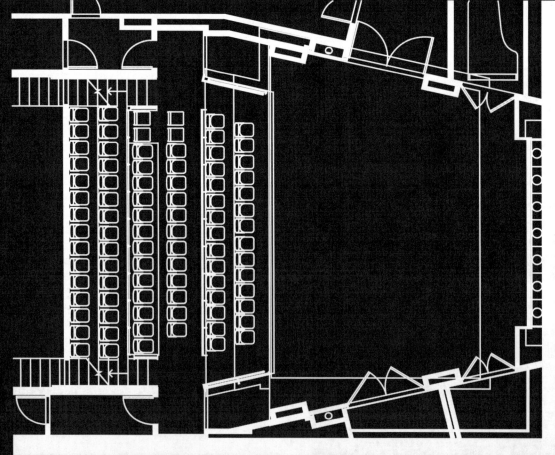

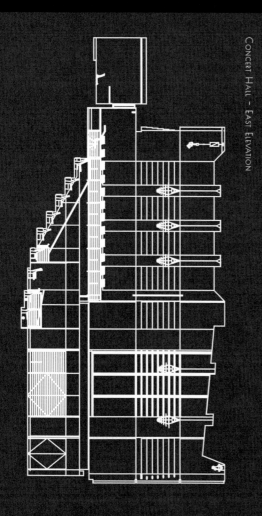

CONCERT HALL - EAST ELEVATION

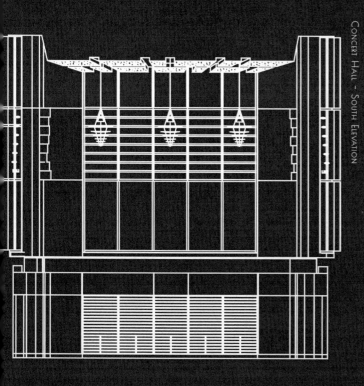

CONCERT HALL - SOUTH ELEVATION

ACOUSTICAL CONSULTANT:
ACENTECH INCORPORATED

ARCHITECT:
ARCHITECTURAL RESOURCES CAMBRIDGE
(AR█

COMPLETION DATE:
2██4

LOCATION:
DEERFIELD, MA I USA

CONSTRUCTION TYPE:
NEW CONSTRUCTION, RENOVATION

CONSTRUCTION/RENOVATION COST:
$33,000,0██

FEATURED SPACE DATA:

ROOM VOLUME:
79,000 F█

FLOOR PLAN AREA:
2,500 F█

SEATING CAPACITY:
16█

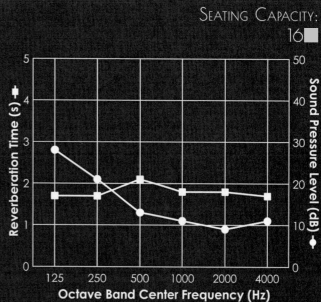

The dramatic arts, musical theater, and music programs at Dr. Phillips High School have a national reputation for excellence. However, their existing auditorium was in need of upgrading and repair to help support their program. Orange County Public Schools approved a Comprehensive Needs Project for the school to provide improved facilities and modernize the campus, including administrative, multi-purpose/dining, classroom, laboratory, art, music suite, media center, science, gymnasium, performing arts center, field house, and football stadium. Renovations included 372,940 gross square feet as well as 35,000 gross square feet of new construction.

The main challenge was to renovate the facility enough to improve the acoustical and theatrical system in the room while staying within the confines of the original building envelope. Another challenge was to make strategic additions to the buildings so that an economical comprehensive series of additions and renovations provided major improvements to the performing arts facility.

Architectural soundscape design methods were utilized by the acoustical consultants to optimize the acoustical, theatrical, and architectural systems in the new and renovated spaces. Discussions were held with performing arts faculty and administrators to assist in developing an acoustical program for the project. The program helped to identify acoustic criteria for each of the rooms. The renovations included the Visual and Performing Arts magnet rooms, such as the auditorium, theater, band classroom, choral

VIEW TOWARD STAGE | STAN KAYE

SIDE WALL | STAN KAYE

MODEL SHAPES OF WALL PANELS | SIEBEIN ASSOCIATES, INC.

classroom, orchestral classroom, ensemble rooms, practice rooms, drama classroom, gymnasium, media center, outdoor covered areas, and student dining room. This was done first by taking a series of impulse response measurements at various locations throughout the existing rooms to use as baseline data for the soundscape discussions.

The highlight of the project is the auditorium/theater space. Acoustic finishes were carefully selected, and analysis was performed to determine their appropriate placement to promote clarity and intelligibility of natural acoustic, reinforced and amplified speech and music. An acoustic throat was formed around the stage to assist with the projection of sounds into the audience. The sidewall reflectors were designed as part of an iterative process with the architect to creatively address the acoustics and aesthetics of the space. The concept for the 3D folded planes was originally developed using origami to stimulate design ideas that could provide sound reflections from the side walls of the room to immerse the audience in a rich, fully-enveloping sound field. Overhead reflectors and wall reflectors were integrated to direct sounds evenly across the audience area. These free-standing forms also reduced the acoustical width of the room and created exit vestibules to reduce sound intrusion from the exterior. Siebein worked closely with the architects to develop custom-angled sidewall reflectors. The final design was a creative collaboration with the architect that evolved from the series of folded paper plane models. These pieces were designed to stand away from the side walls and propagate sound across the seating area. They included an angled shelf piece at the top that helps direct reflections back into the audience area. These freestanding elements also doubled as elements to house the acoustic drapes for the side walls when not in use. The background noise level (HVAC system on) and reverberation time data were measured in the unoccupied space.

The results of this effort are a modern, sleek, up-to-date auditorium and other magnet program spaces that support the caliber of the fine arts program at this very special school.

Opposite page: View from Stage | Stan Kaye

DR. PHILLIPS HIGH SCHOOL

Section 1 C.T. Hsu + Associates, P.A.

1. Convex-curved Reflecting Ceiling Panels 2. Convex-curved Reflecting / Diffusing Wall Panels

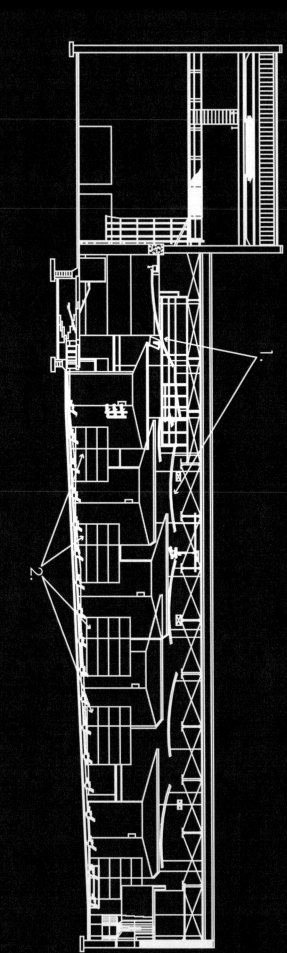

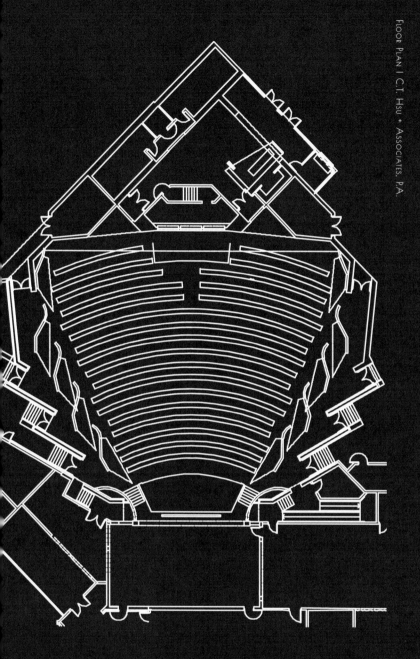

Floor Plan | C.T. Hsu + Associates, P.A.

ACOUSTICAL CONSULTANT:
SIEBEIN ASSOCIATES, INC.

ARCHITECT:
C.T. HSU + ASSOCIATES, P.A.

COMPLETION DATE:
2013

LOCATION:
ORLANDO, FL | USA

CONSTRUCTION TYPE:
RENOVATION

CONSTRUCTION/RENOVATION COST:
$6,660,000 (CAMPUS-WIDE)

FEATURED SPACE DATA:

ROOM VOLUME:
179,920 FT3

FLOOR PLAN AREA:
7,214 FT2

SEATING CAPACITY:
696

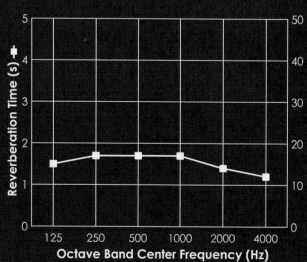

Edgewater High School is one of Orlando's oldest and most visible schools that was to be transformed to be the prototype modern urban design campus with innovative and state-of-the-art design methods and solutions for architectural systems for the theater, band room, choral room, orchestra room, ensemble rooms, practice rooms, gymnasium, atrium, cafeteria, and library. The acoustical finishes in each space were specially designed to reflect the dynamic, modern aesthetic design of the building.

The 422,000 square feet high school was designed with a three-story tall glazed atrium that forms the 'spine' of the school. This atrium becomes the grand circulation space that opens into a large day-lit cafeteria gathering space. Wings of educational and administrative spaces become the joints of the interactive multi-level concourse.

The new 750-seat O. R. Davis Auditorium is a multi-purpose performing arts theater and is the icon of the campus: greeting students, staff, and parents, and reflecting the culmination of the dedication and talent of the students and instructors of the school. The theater was acoustically and visually designed to propagate sounds to the entire audience with both clarity and reverberance with strategic shaping of the ceilings and walls. The concept of the modern dynamics of the school with vibrant colors and materials, architectural sculpting of interior and exterior surfaces, and a visual itinerary that forms the aesthetic intention of the entire school is also

activated in the theater.

Wood panel framing of the proscenium opening is encased with convex-curved shaped walls and ceiling that forms the 'acoustic throat' of the theater that propagates sound reflections from multiple sources located on the stage throughout the entire audience area with early sound reflections. The ceiling with suspended convex-curved panels are strategically radiused and angled so that sounds are progressively reflected off the panels to evenly distribute loud and clear natural and amplified sounds down to the audience. The custom 'origami' panels on the side walls accentuate the modern aesthetics. These panels are specifically shaped to provide diffuse lateral sound reflections towards the center seating areas to enhance the sense of envelopment and intimacy in a room that is visually grand. Sound absorbent panels of varying thickness are located on the rear walls to balance the reverberance of the room to accommodate a variety of venue types ranging from classical music performances, theatrical productions, audio visual presentations, and lectures.

Manufactured sound diffusing and absorbing materials were strategically designed to provide both aurally enhanced sound fields and visually aesthetic environments for instructors and students in the music rooms. Balancing costs, timing, and aesthetics, the music rooms, in particular, are acoustically treated with sound absorbing panels and sound diffusing panels that are arranged as artistic patterns taking advantage of varying sizes of panels to create

AUDITORIUM – SIDE WALL | SIEBEIN ASSOCIATES, INC.

CLASSROOM | SIEBEIN ASSOCIATES, INC.

harmonized motifs reflective of the talented students and instructors that use the room day by day. The motifs vary from music room to music room to reduce excessive loudness of percussion and brass instruments while allowing the subtleties of flutes and violins to be enhanced.

Acoustical design permeates all of the buildings on campus. Reverberant noise reduction was achieved through careful integration of visually dynamic, durable acoustical treatment on the ceilings, walls, and windows of the atrium, gymnasium, cafeteria, classrooms, and corridors. Sound isolating assemblies were selected and constructed to reduce sound bleed between music rooms and between music rooms and adjacent classrooms to meet or exceed the latest acoustical design criteria for educational spaces.

The school was transformed into a visually and acoustically dynamic school that exceeds the latest acoustical design guidelines for performance and music rehearsal spaces. Close collaboration with the design team resulted in acoustically enhanced and dramatic spaces that house a magnet program in the performing arts. Visually dynamic integration of economical acoustical shaping and finishes encouraged students and staff to take pride in the richness of the visual and aural experience of the newly remodeled school.

The atrium is the first multi-level commons area in the entire school district, where large groups of students can gather and talk in a comfortable acoustical environment while integrating shaped perforated metal sound absorbing panels throughout the space that creates the modern aesthetics throughout the concourse. The integration of daylighting and acoustical materials through sculpted architectural forms provide a social learning environment that invites active participation from students as members of a community. The two-story tall cafeteria is also a part of the atrium which required acoustical treatment that is integrated into not only the wall surfaces but also custom acoustical baffles on the mullions of windows to reduce excessive noise levels from dining activities and large groups of students talking within the dining area as well as to reduce noise from propagating throughout the atrium.

In addition to the architect and acoustical consultant, the design team included Gilbane Building Company (general contractor).

Above: Auditorium | Siebein Associates, Inc.

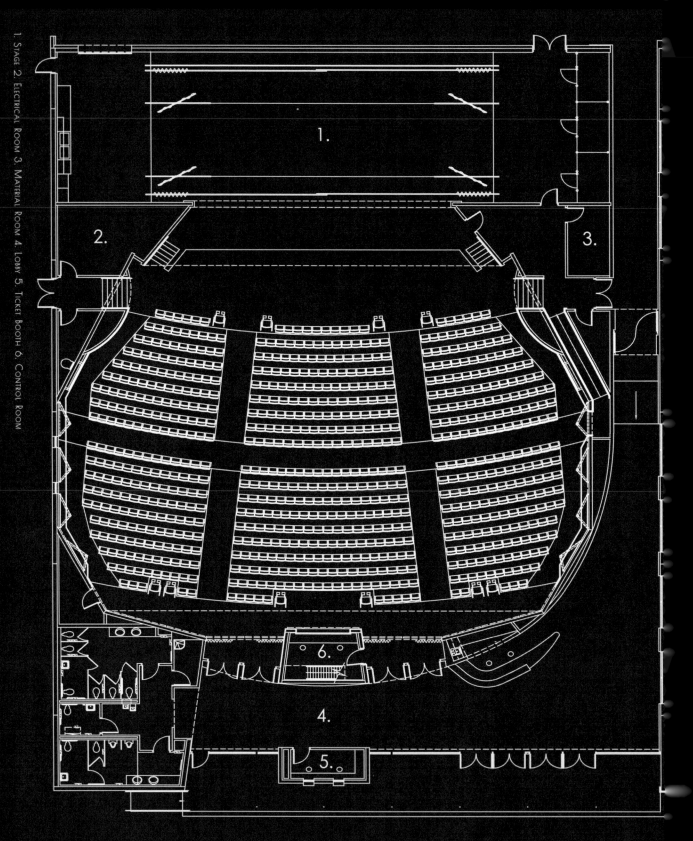

ACOUSTICAL CONSULTANT:
SIEBEIN ASSOCIATES, INC.

ARCHITECT:
C.T. HSU + ASSOCIATES, P.A.

COMPLETION DATE:
2012

LOCATION:
ORLANDO, FL | USA

CONSTRUCTION TYPE:
NEW CONSTRUCTION

CONSTRUCTION/RENOVATION COST:
$57,000,000

FEATURED SPACE DATA:

ROOM VOLUME:
287,090 FT3

FLOOR PLAN AREA:
21,892 FT2

SEATING CAPACITY:
750

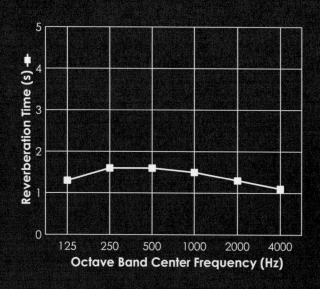

By 2015, New Trier High School's student body of 4,000 had outgrown its 117-year old campus. With no available land on which to construct new facilities, the school's existing cafeteria, tech arts, and music buildings were demolished to allow a new, 4-story structure to be nestled between the remaining classroom, gymnasium, and auditorium theatre buildings. The new LEED-Gold certified addition houses a library, cafeteria, applied arts/maker spaces, a culinary lab, interconnected visual arts studios, a technology support bar, STEM laboratories, core and flexible classrooms, and the school's performing arts facilities. The performing arts spaces include orchestra, band, jazz, chamber and two choral rehearsal rooms with the larger of the two choral rooms doubling as a recital and dance rehearsal space. A music theory and keyboarding lab,

proscenium theatre, studio theatre, drama rehearsal, recording studio, media stage/video studio and a radio station round out the rooms supporting a robust performing arts curriculum.

The spaces within New Trier's performing arts wing are continuously programmed throughout the school day and well into the evening hours, placing elevated performance requirements on the isolation systems employed in the compact, steel-framed building. The rehearsal rooms employ two distinct sound isolation techniques based on the sound power levels of the activities within, their location within the building, and the predetermined structural spans.

The orchestra, band, and jazz rehearsal spaces, sitting above the theatre shop and below science laboratories, utilize a box-in-box construction with grout filled,

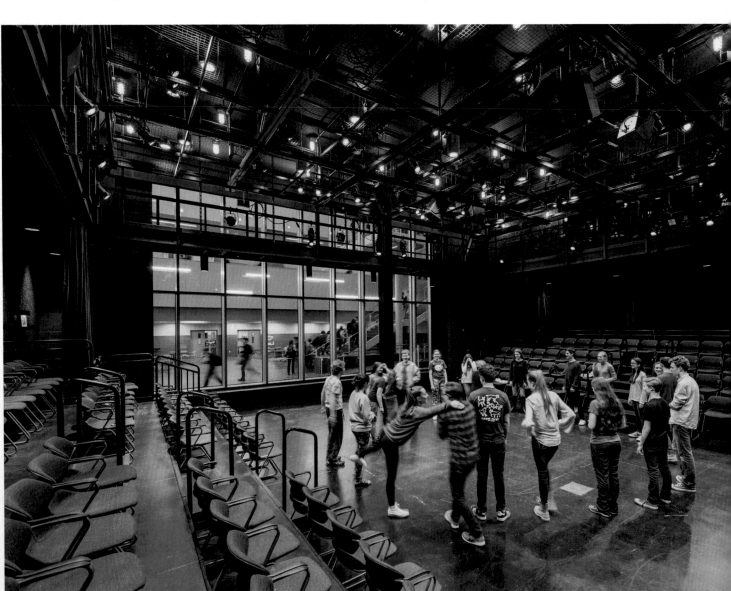

split-face block forming the interior perimeter of the rooms. The orchestra and jazz rooms' block walls rest on the concrete structural slab with fiberglass pads supporting a floating wood floor system. As the potential of incorporating a structural break in the concrete slab between the three adjacent rehearsal spaces was not an option, the band room, which is sandwiched between the orchestra and jazz rooms, utilizes a jack up concrete slab, with spring isolators supporting both the grout filled block walls and the floor system. All three rehearsal rooms include spring/neoprene-hung gypsum barrier ceilings.

Concrete and block isolation systems were not an option for the choral rooms, as the long structural spans required for the 200-seat proscenium theatre below restricted the building loading. The resilient inner boxes of the choral rooms were formed with a total of 8 layers (5+3) of 2.2-pound-density gypsum board wall systems on separate studs resting on resilient fiberglass matts. These spaces are also used for musical theatre rehearsal, imposing rhythmic dancing loads that could not interrupt performances in the proscenium theatre below. With predetermined floor-to-floor heights required to align new floors with existing ones in the adjacent building, every inch of available height was required in the theatre to reach the desired seat count and lighting angles from catwalk positions. Even with the use of resilient floor and ceiling systems, the natural frequency of the long span structure would have been too low to mitigate structural vibration induced by the rhythmic impact of dance. To sufficiently raise the structural floor's natural frequency, columns were added within the theatre's audience chamber to reduce the span.

Approaches to rehearsal room acoustics varied with ensemble type and size as well as location within the building. Choral rooms were constructed of gypsum board to keep partition weights to a minimum. Non-orthogonal floor plans allowed flutter control without special surface treatments, keeping costs to a minimum. On the ground floor, a superimposed column grid and large ensemble sizes limited the ability to angle the wall surfaces for the orchestra, band, and jazz rooms. A combination of split-face block, tuned absorption/diffusion panels, fabric-wrapped absorption panels, and offset locker faces lining the perimeter of the rehearsal rooms control both reflection patterns and sound levels.

In addition to the architect and acoustical consultant, the design team included: Schuler Shook (theatre consultant), Threshold Acoustics LLC (audio systems designer, video systems designer), Wight & Company (mechanical engineer), and Pepper Construction (general contractor). The reverberation time data shown were calculated for the unoccupied space. The background noise level data were measured in the unoccupied space with the HVAC system on, operating at full occupancy loading.

Opposite page: Studio Theatre | Steinkamp Photography

THEATRE | STEINKAMP PHOTOGRAPHY

LARGE CHORAL REHEARSAL ROOM | STEINKAMP PHOTOGRAPHY

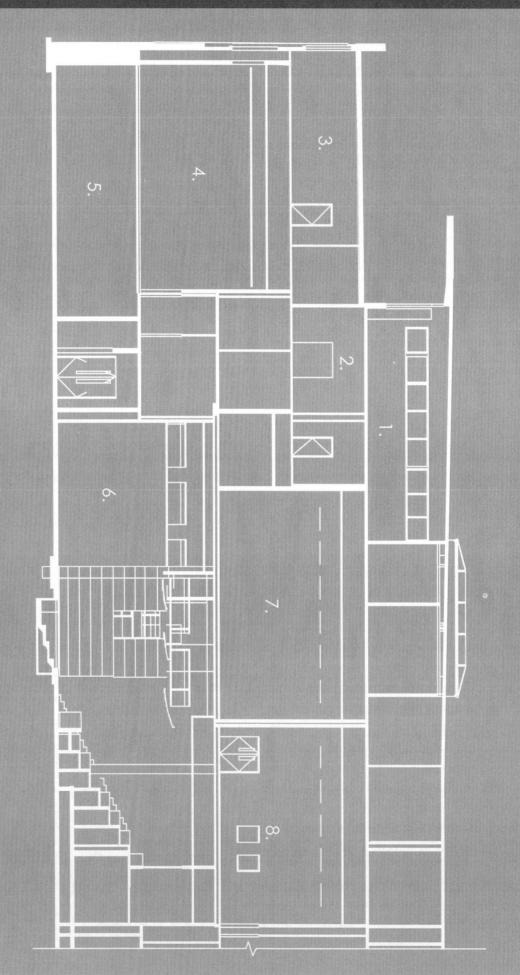

1. Class Studio 2. Science Prep 3. Science Lab 4. Band Rehearsal 5. Theatre Shop 6. Theatre 7. Large Choral Rehearsal and Recital 8. Small Choral Rehearsal

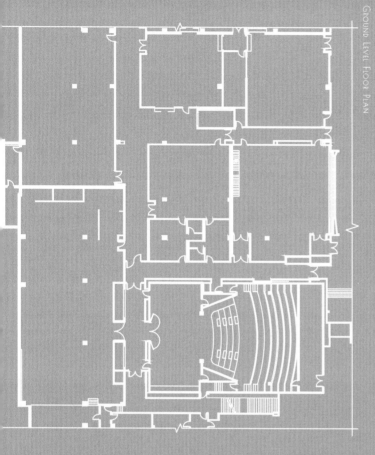

GROUND LEVEL FLOOR PLAN

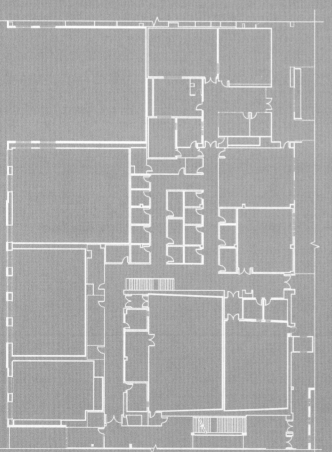

SECOND LEVEL FLOOR PLAN

ACOUSTICAL CONSULTANT:
THRESHOLD ACOUSTICS LLC

ARCHITECT:
WIGHT & COMPANY

COMPLETION DATE:
2018

LOCATION:
WINNETKA, IL | USA

CONSTRUCTION TYPE:
NEW CONSTRUCTION

CONSTRUCTION/RENOVATION COST:
$82,000,000

FEATURED SPACE DATA:

ROOM VOLUME:
50,400 FT3

FLOOR PLAN AREA:
2,200 FT2

SEATING CAPACITY:
65

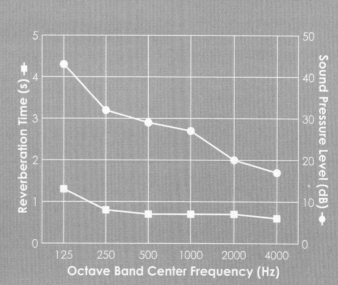

Phillips Exeter Academy is a private school for grades 9 through 12 with a strong arts program. Students come from 43 states and 29 countries. The first addition to the 1950s Forrestal-Bowld Music Center, opening in 1996, included a gracious rehearsal room for orchestra and band, plus practice rooms and teaching studios. The second addition, completed in 2016, included ensemble rehearsal rooms, classrooms, and teaching studios, as well as a small recording studio and electronic music spaces. The overall programming goal was to provide a well-equipped, multi-faceted facility that can enrich the lives of both students and community with a rich and diverse set of musical offerings.

At the heart of the newest addition is "The Bowld", a soaring rehearsal/recital space that can accommodate a full symphony orchestra while also providing a recital space for the school's students and faculty. The room has flexible seating, lighting and adjustable acoustic curtains that cater to the full range of programs. The space also houses audio visual systems featuring a projector and screen, loudspeakers lowered from the ceiling for light reinforcement, and paired ceiling mics for recording of rehearsals and performances.

The Recital Studio appears a simple room with a glass wall to the north and a bank of raised bench seating to the south. With the addition of loose chairs, the room can accommodate an audience of 300 for recitals. Subtle, projective shaping of the rooms side walls provide support for recitalists while preventing large ensembles from becoming uncomfortably loud. Wood grilles ring the upper volume of the

room, concealing various sound-reflecting, sound-absorbing, and sound-diffusing surfaces, a set of curtains pockets, and the acoustic curtains. A sweeping reflector overhead allows the room to serve both its rehearsal and recital functions without compromise. A large sliding door allows audiences to flow in from the lobby before and after an event, while minimizing distraction during an event.

As a recital hall the room sounds spacious with a lovely blend of clarity and reverberance served by the careful shaping of sound sustaining surfaces, the placement of finishes and the gracious height of the space. The use of variable acoustic finishes gives the room considerable flexibility: sun shades and banners can be exposed into the room volume as necessary providing a wide range in reverberation times for the space. The reverberation time data shown were measured in the unoccupied space iwth curtains retracted. These versatile finishes allow for user-control of the build-up and decay of sound in the space making the Recital Studio suitable for everything from choral to jazz.

In addition to the architect and acoustical consultant, the design team included: Martin Vinik Planning for the Arts (theatre consultant), Kirkegaard (audio systems designer, video systems designer), and Consigli Construction Co (general contractor).

Opposite page: Rehearsal Hall - Interior View of Stage

Above: Rehearsal Hall - Audience

REHEARSAL HALL - ENTRANCE

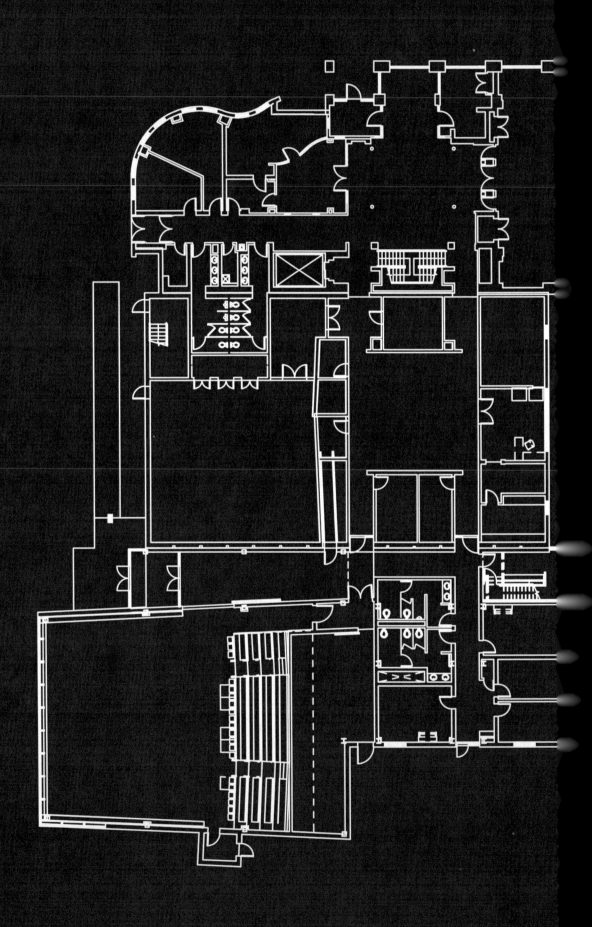

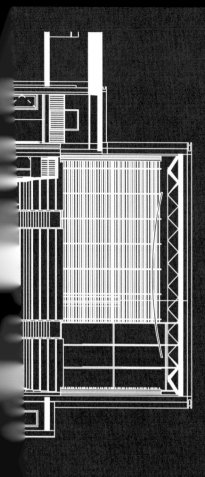

Rehearsal Hall - Cross Section

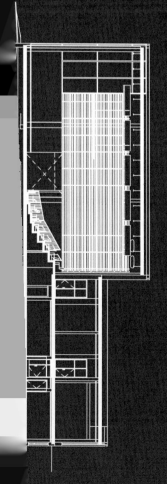

Rehearsal Hall - Longitudinal Section

ACOUSTICAL CONSULTANT:
KIRKEGAARD

ARCHITECT:
WILLIAM RAWN ASSOCIATES

COMPLETION DATE:
2016

LOCATION:
EXETER, NH I USA

CONSTRUCTION TYPE:
NEW CONSTRUCTION, RENOVATION,
EXPANSION

CONSTRUCTION/RENOVATION COST:
$10,100,000

FEATURED SPACE DATA:

ROOM VOLUME:
119,000 FT3

FLOOR PLAN AREA:
4,300 FT2

SEATING CAPACITY:
300

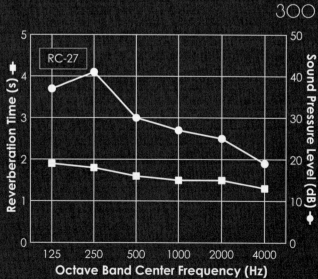

The $36 million Rainey-McCullers School of the Arts is a magnet school designed to accommodate 500 students in grades 6-12. The three-story, 118,500-square-foot facility in Columbus, Georgia is named after famous Columbus natives Ma Rainey, a blues singer, and Carson McCullers, a novelist. Rainey-McCullers is an audition based, performing and visual arts magnet school for students living in Muscogee County, Georgia that offers programs in music (band, chorus, orchestra, classical guitar, and piano), dance (classical and modern), theater, technical theater, musical theater, visual art (2-D, 3-D, sculpture and photography), film and creative writing. Opened in 2017, the school includes a 650-seat state-of-the-art auditorium, dedicated rehearsal spaces for band, orchestra and chorus, a piano lab, practice rooms for individual students and music ensembles, a black box theater, two dance studios, a recording studio suite, a digital editing room, film screening room, darkroom, art classrooms, an art gallery, and general education classrooms.

The aesthetic concept established during schematic design featured the use of wood finishes arranged in wave-like, undulating gestures to compose the shaping of the Theatre. Curved, wood-veneer panels were selected for the overhead ceiling reflectors and a series of undulating wood-veneer panels were used to create the acoustical throat and the side wall shaping in the space. To ensure that the curved elements would function acoustically as well as aesthetically, an iterative ray tracing study was conducted to determine what curve radii would provide the best series of reflections from the side wall reflections to enhance the spaciousness and envelopment in the room as well as advantageous overhead reflections to enhance the intimacy and clarity of sounds in the space. This was particularly important for side wall panel elements because of the presence of convex-curves to form the wave-like gesture. Because the Theatre is used for varying performance types including both natural acoustic performances and performances with amplified speech and music, the space was designed to have a reverberation time of approximately 1.1 to 1.3 seconds in the mid-frequencies to provide speech clarity for theatrical events, lectures, films, and ceremonies. The measured impulse in the room shows early reflections from the ceiling and side walls that enhance the natural acoustic performances, as well as even decay. To maintain the curved wood panel aesthetic, the panels located along the back wall were perforated with sound absorbent material provided behind the panels.

The dedicated band, orchestra, and choral rehearsal spaces are high ceilinged spaces to provide the appropriate volume for the full sound of large music ensembles. Each space is carefully tuned with a combination of sound absorbent treatment to control the reverberation in the spaces to promote the listening clarity of each note while evenly distributed sound diffusion on the ceiling and wall surfaces allow the members of each musical ensemble to hear

THEATRE | HECHT BURDESHAW ARCHITECTS, INC.

THEATRE PARTERRE VIEW | HECHT BURDESHAW ARCHITECTS, INC.

each other during rehearsals. Containing the most absorption to control the sound levels produced by brass and percussion instruments, the reverberation times measured in the Band Room were 0.5 to 0.6 seconds in the mid-frequencies. The mid-frequency reverberation times measured in the Orchestra Room were 0.7 seconds in the mid-frequencies. Containing the least amount of absorption to help support the presence of the vocal output of the students, the mid-frequency reverberation times measured in the Chorus Room were 0.7 to 0.8 seconds in the mid-frequencies. Each of the music rehearsal spaces is located amongst buffer spaces including storage rooms, music teacher offices, and corridors to reduce the critical adjacencies between learning spaces. STC rated doors were used at the entrances to each space.

The new Rainey-McCullers School of the Arts also provides students with the opportunity to record music and voice-overs and learn to edit the recorded content with a state-of-the-art recording studio suite. The interior walls and ceiling of the Recording Studio Control Booth were shaped to avoid room modes and provide a reflection free zone at the mix and listening location. This combined with field fabricated corner bass traps helped to reduce the low frequency reverberation time. The Control Room is connected to an Isolation Booth by a sound lock vestibule. The Isolation Booth contains room for a small number of individuals to record within a sound absorbent room. An STC-rated window in the Isolation Booth allows for a visual connection back into the Control Room and to bring more light into the Isolation Booth making the space feel larger.

The Black Box theatre is designed to maximize the versatility of the space to accommodate a variety of performance types with flexible configurations and playing directions. With a combination of diffusing and absorbent panels evenly spread throughout the space, the Black Box Theatre may be used for dramatic performances, poetry readings, and experimental theater as well as for music practice and performances. Outfitted with full-height retractable acoustical drapes that can be deployed along all four walls, the Black Box can also be deadened for the spoken word and theatrical productions.

In addition to room acoustic design, the design of noise and vibration control systems for the HVAC system and sound isolation system serving the music classrooms, recording studio suite, performance spaces, and other acoustically sensitive spaces throughout the school was undertaken.

In addition to the architect and acoustical consultant, the design team included: Theatre Consultants Collaborative, Inc. (theatre consultant), Peach Engineering (mechanical engineer), and Brasfield & Gorrie (general contractor). The reverberation time data shown were measured in the unoccupied space. The background noise level data shown were measured in the unoccupied space with the HVAC system on.

Orchestra Room | Hecht Burdeshaw Architects, Inc.

Piano Lab | Hecht Burdeshaw Architects, Inc.

Floor Plan.

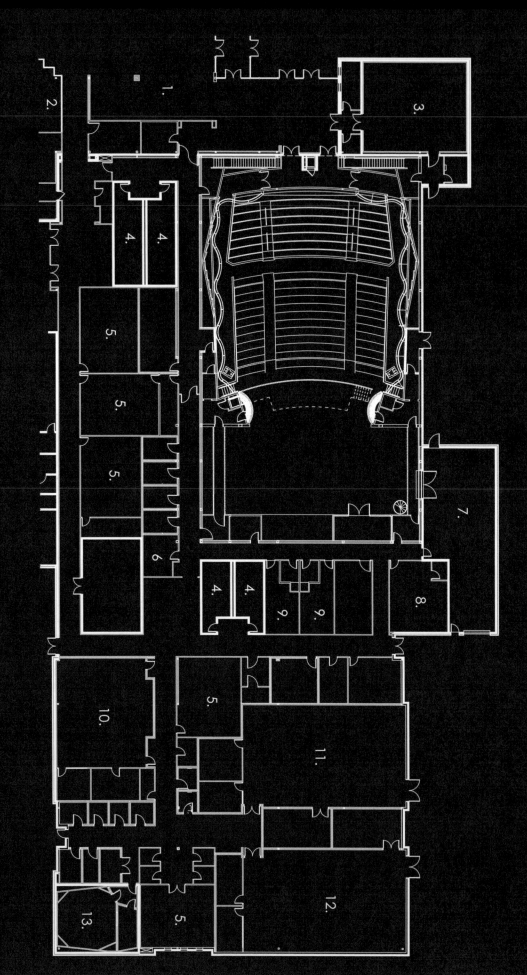

1. Gallery 2. Media Center 3. Black Box 4. Restrooms 5. Classrooms 6. Green Room 7. Shop/Prep Storage 8. Costume Storage 9. Dressing Rooms 10. Choral 11. Band 12. Orchestra 13. Recording Studio Suite

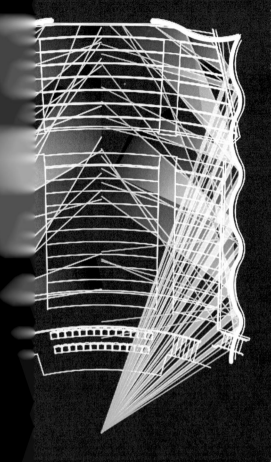

PLAN RAY TRACE

SECTION RAY TRACE

ACOUSTICAL CONSULTANT:
SIEBEIN ASSOCIATES, INC.

ARCHITECT:
HECHT BURDESHAW ARCHITECTS, INC.

COMPLETION DATE:
2017

LOCATION:
COLUMBUS, GA I USA

CONSTRUCTION TYPE:
NEW CONSTRUCTION

CONSTRUCTION/RENOVATION COST:
$36,000,000

FEATURED SPACE DATA:

ROOM VOLUME:
240,000 FT3

FLOOR PLAN AREA:
8,400 FT2

SEATING CAPACITY:
650

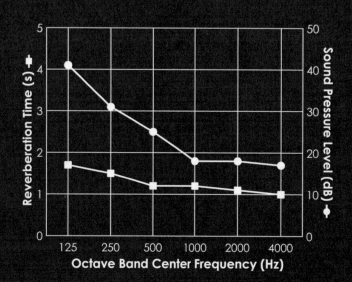

The Carter Center for Music Education & Performance houses the educational arm of the Rhode Island Philharmonic, providing rehearsal space for full orchestra, wind ensemble, sectionals, and individual practice. There are 1600 students enrolled in the music school, served by 70 faculty members. The merger between the Rhode Island Philharmonic and the Music School allows the Philharmonic to satisfy its educational mission while cultivating an early appreciation for classical music events.

The building structure was previously used for office and light manufacturing. After significant renovations, the 31,000-square foot facility contains 31 private rehearsal studios, classrooms for early childhood education and music therapy, and two large rehearsal halls for student and community orchestras. The larger ensemble spaces are sufficiently sized for the Rhode Island Philharmonic and other orchestras to rehearse and make recordings. School concerts are often held in the rehearsal halls, seating up to 200 patrons. A recording studio and associated control booth are also part of the renovation.

Throughout the facility, the music rooms feature double stud walls and double ceilings. The finished ceiling is acoustic tile to control reverberation with a gypsum board ceilings above to provide sound isolation. Each music room also includes areas of acoustic wall panels to control reverberation and prevent flutter echoes. The Suzuki Classroom and Early Childhood Classroom also include movable drapery, so instructors can adjust the sound

absorption to their preference. A raised isolation floor (atop the existing structural floor) is incorporated at the Control Room and Recording Studio to provide greater control of flanking sound in these rooms.

A separate wing is designed for jazz, rock, and blues. This wing includes dedicated spaces for smaller ensembles and is designed with more sound absorptive materials to provide the shorter reverberation times necessary for these uses. Placing these rooms in a separate wing of the building minimizes sound transfer from the louder, amplified music to the more sensitive orchestra rehearsal rooms, as well as the recording studio.

Acoustic features in the large and medium rehearsal rooms include sound-diffusing elements on the walls and ceilings. Cavanaugh Tocci worked with KITE

Architects to design and integrate cost-effective acoustical treatments into the wall structures of the rooms as well as custom diffusive elements that could be constructed on site. Both rooms included exposed structure for maximum height and interior room volume with suspended clouds for diffusion and early reflections.

In the large rehearsal room, diffusion is provided by serpentine contours constructed of gypsum board. The sidewalls have wood paneling in a largescale basket-weave pattern to provide surface relief for lateral reflections. Similar concepts are used in the medium rehearsal room, in the form of angular undulations of the walls and V shaped overhead reflectors.

Designing a high-quality community music center given constraints of existing building structures and modest budgets can present many acoustical challenges. As acoustical consultants, these challenges are rarely overcome through our own efforts alone. David Beauchesne, the executive director of the Rhode Island Philharmonic Orchestra & Music School, provided a clear vision of the project goals and insisted that acoustics remain a priority throughout design and construction. Through the diligent work and creativity of the board of directors, design team, and construction manager, the Carter Center for Music Education & Performance is a resounding success.

In addition to the architect and acoustical consultant, the design team included E.W. Burman, Inc. (general contractor). The reverberation time data were measured in the unoccupied space, and the background noise level data were measured in the unoccupied space with the HVAC system on.

Opposite page: View of Medium Rehearsal Room | Cavanaugh Tocci

VIEW OF LARGE REHEARSAL ROOM

1 Studio 2 Keyboard Studio 3 Medium Rehearsal 4 Large Rehearsal 5 Rehearsal

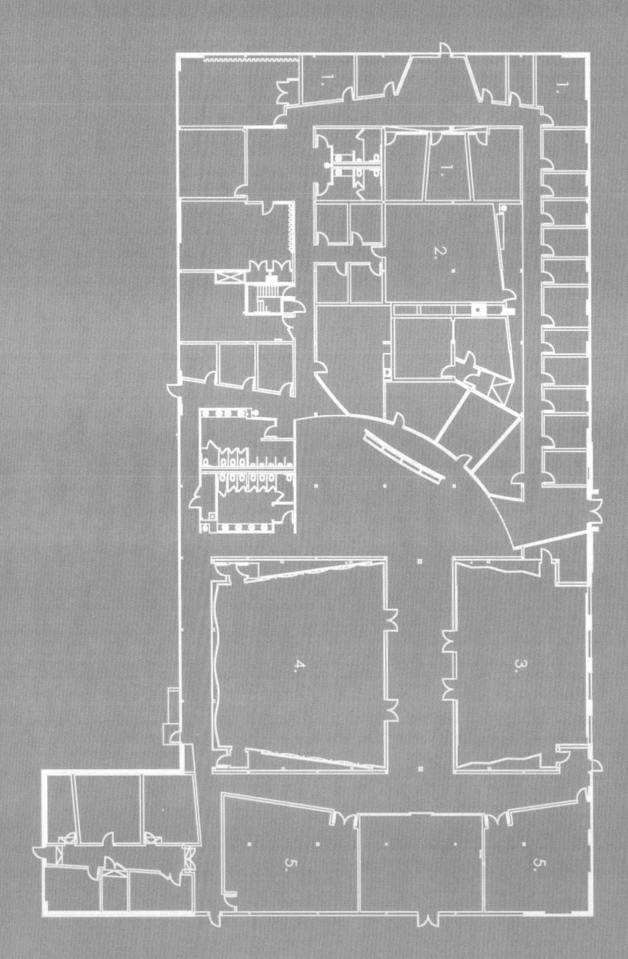

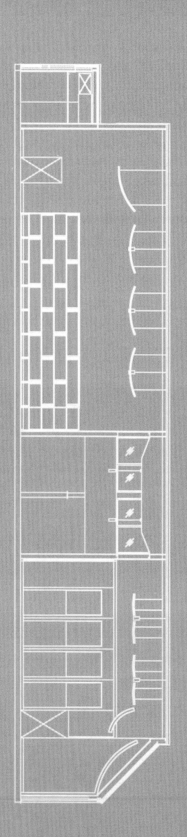

LONGITUDINAL SECTION

ACOUSTICAL CONSULTANT:
CAVANAUGH TOCCI ASSOCIATES, INC.

ARCHITECT:
KITE ARCHITECTS

COMPLETION DATE:
2008

LOCATION:
EAST PROVIDENCE, RI I USA

CONSTRUCTION TYPE:
RENOVATION

CONSTRUCTION/RENOVATION COST:
$8,000,000

FEATURED SPACE DATA:

ROOM VOLUME:
60,600 FT3

FLOOR PLAN AREA:
2,550 FT2

SEATING CAPACITY:
200

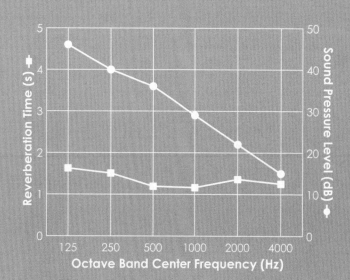

The Black Box Theater and Instrumental Room of the Trinity Preparatory School were part of a significant renovation/addition project to accommodate the growing needs of the performing arts program of the 6th through 12th grade premiere college preparatory school. The school has an active music, dance and experimental theater program with highly acclaimed faculty who teach classical, jazz, choral, and other contemporary large and small ensembles and acting. The space had to provide acoustical qualities including texture, clarity, warmth, envelopment, and a calculated amount of reverberation for critical listening by students and instructors as well as for students to hear each other during rehearsals.

The Black Box Theater is a multi-purpose space with windows that overlook the lawn of the campus. It was designed to be used for a variety of performances and rehearsals including experimental plays, dance and music rehearsals and recitals. The space is also used for lectures with and without audio visual presentations, poetry readings, student meetings, cheerleader rehearsals, physical education activities and much more. The acoustical design for a multi-purpose space where students and teachers can develop their skills and excel in the arts was a challenging and rewarding experience.

Accommodating such a variety of uses requires special space-shaping and acoustical finishes that can be configured differently according to the size and type of ensemble, performance or meeting.

Suspended sound diffusing ceiling panels that are

Black Box Windows | Siebein Associates, Inc.

Black Box Shaped Walls | Siebein Associates, Inc.

Ceiling System | Siebein Associates, Inc.

Instrumental Room Before Renovation | None

integrated above the lighting grid system provide diffuse propagation of sounds so that performers can hear themselves as well as to enhance the presence of natural acoustic sounds during music performances and recitals. The shaped walls provide envelopment of diffuse sound reflections throughout the room when the acoustical drapes are retracted. Retractable acoustical drapes were integrated on three walls to enhance clarity of speech and to vary reverberance in the space. Sound isolating ceiling, wall, and floor assemblies were implemented to reduce audible transmission of sounds between the black box theater and the adjacent second floor classroom. To enhance the natural and amplified sounds, the mechanical system was designed to provide a quiet environment where every note or syllable can be heard. Initial post-construction measurements conducted showed

that the background noise criteria was not met due to duct borne noise from the operating unit and noise generated by high air-velocities. Therefore, the acoustical consultants collaborated with the architect and mechanical engineer to fine-tune the background noise levels in the space. As a result, ducts were re-sized so that supply and return air velocities meet recommended guidelines and sound attenuators were installed near the mechanical unit.

Both the staff, students and their parents were elated to have rehearsal and recital spaces that are used every day with joy for the growing performing arts program.

The existing instrumental room was enlarged significantly both vertically and horizontally to accommodate larger ensembles as well as to provide an acoustical volume for the dissipation of high energy sounds and room geometry to enhance overhead sound reflections.

Field fabricated convex-curved sound diffusing ceiling and wall panels were constructed to provide diffuse and clear sound reflections so that both students and instructors may hear well. Sound absorbent materials were strategically installed to reduce acoustical defects such as prolonged reverberation and harsh sound reflections that were present in the previous environment. Acoustical drapes that could be retracted into pocket walls were integrated into the room design so the acoustics could be adjusted to accommodate the varying sizes of ensembles and genres of music rehearsed in the room.

In addition to the architect and acoustical consultant, the design team included: TLC Engineering for Architecture (mechanical engineer), Brasfield & Gorrie (general contractor), Bishop Engineering Company (structural engineer), and TLC Engineering for Architecture (electrical engineer). The reverberation time data shown were calculated for the unoccupied space. The background noise level data shown were measured in the unoccupied space with the air-conditioning system operating at full cooling capacity.

Instrumental Room After Renovation | Siebein Associates, Inc.

Acoustical Drape Pockets | Siebein Associates, Inc.

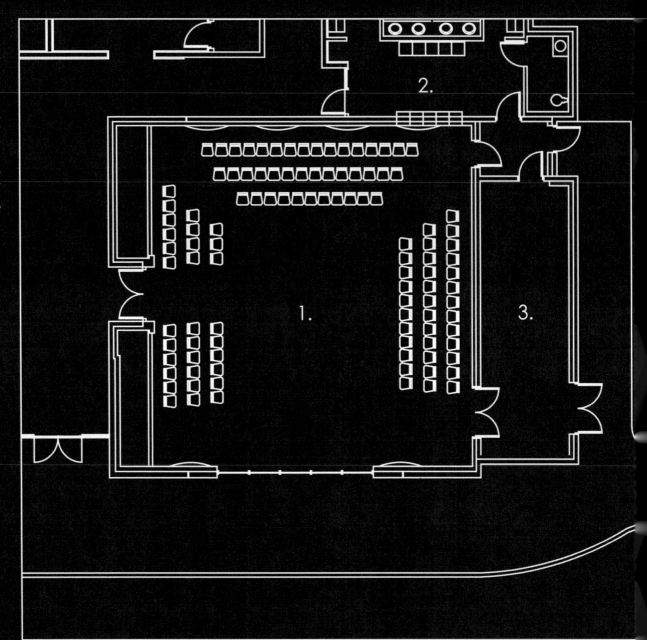

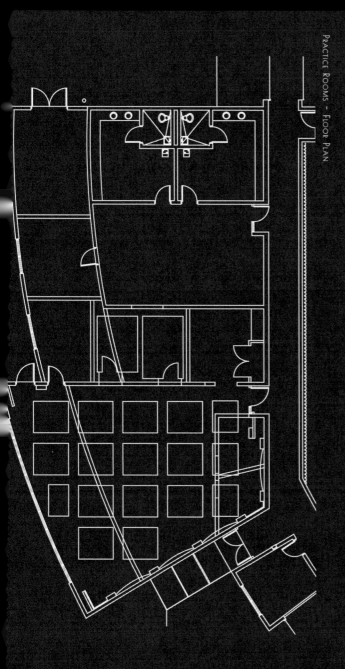

PRACTICE ROOMS - FLOOR PLAN

ACOUSTICAL CONSULTANT:
SIEBEIN ASSOCIATES, INC.

ARCHITECT:
A.D. HETZER ARCHITECTS

COMPLETION DATE:
2016

LOCATION:
WINTER PARK, FL I USA

CONSTRUCTION TYPE:
NEW CONSTRUCTION, RENOVATION

CONSTRUCTION/RENOVATION COST:
$5,200,000

FEATURED SPACE DATA:

ROOM VOLUME:
52,173 FT3

FLOOR PLAN AREA:
1,705 FT2

SEATING CAPACITY:
122

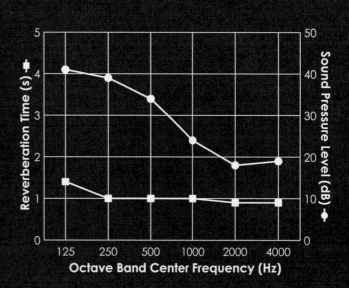

The Union Public School's music program begins at the sixth-grade level with an annual enrollment of approximately 400 students. Music offerings include marching band, concert bands, pep bands, jazz bands, and orchestras. The orchestra program has a reputation for excellence in the state, region, and nation with numerous awards at each of these levels. Union was one of two Oklahoma districts recognized by the National Association of Music Merchants (NAMM) Foundation as Best Communities in Music Education for their outstanding music education programs.

Acoustonica, LLC provided acoustical consulting services for the new construction of their fine arts addition. The project involved complete renovation of the existing music facility and the addition of three band rooms, three orchestra rooms, and a total of five practice rooms. The new addition is linked to the main Fine Arts Building. The project also included choral rehearsal rooms and faculty offices/studios. The additions totaled approximately 17,000 square feet. In addition to serving as the acoustical consultant, Acoustonica, LLC was the theatre consultant and audio systems designer for the project.

The mid-frequency reverberation times in the orchestra rooms averaged from 0.65-0.70 seconds and were calculated for the unoccupied spaces. The music rooms have carpet tiles on the floors, and the wall construction between studios consists of double steel stud partitions with a total Sound Transmission Class (STC) rating of 68. The ceilings in the music rooms consisted of typical lay-in ceiling tile with a

Noise Reduction Coefficient (NRC) of 0.90. Pyramid diffusers are placed strategically to diffuse sound in the spaces. Walls received sound-absorbing treatment with fabric-wrapped acoustical panels. The panels were placed above head height to help accommodate storage units for the musical instruments.

The practice rooms were placed in a cross layout with four rooms sharing common walls. The wall construction between practice rooms consists of double stud gypsum board with an STC rating of 65. The walls were detailed so that no practice room would have a structural connection with another practice room. This design helped achieve the required performance and helped control structure-borne noise from transferring between practice rooms.

The mechanical system was designed to achieve a Noise Criteria rating of NC 30. The noise ratings were calculated for the unoccupied space, and the background noise level data were measured in the unoccupied space with the HVAC system on. The rooftop units were placed above spaces not sensitive to noise to help reduce noise and vibration impact and to allow enough duct length for smooth and quiet air flow.

Traffic noise was also considered in the design. Exterior sound measurements were recorded to help establish the existing background noise levels. The data were used to help calculate the required noise reduction through the building façade and helped with the selection of the exterior glass.

Opposite page: Band Room | Walid Tikriti / Acoustonica

Above: Band Room Alternate View | Walid Tikriti / Acoustonica

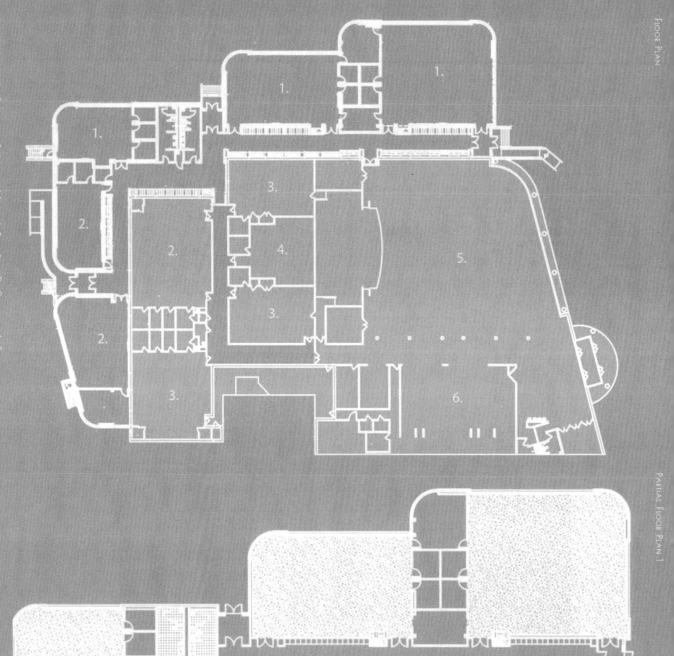

1. BAND ROOM 2. ORCHESTRA 3. VOCAL MUSIC 4. FINE ARTS 5. CAFETERIA 6. SERVERY

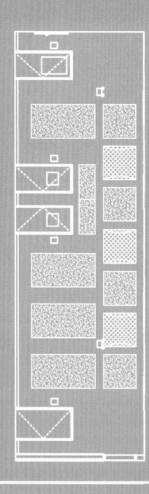

BAND ROOM - WEST ELEVATION

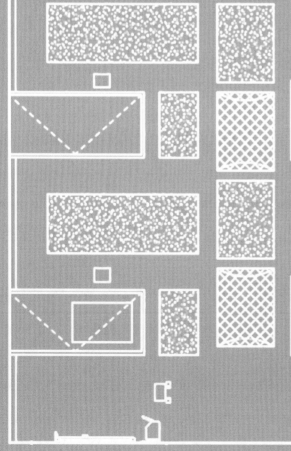

ORCHESTRA ROOM - NORTH ELEVATION

ACOUSTICAL CONSULTANT:
ACOUSTONICA, LLC

ARCHITECT:
PSA DEWBERRY

COMPLETION DATE:
2013

LOCATION:
TULSA, OK | USA

CONSTRUCTION TYPE:
EXPANSION

CONSTRUCTION/RENOVATION COST:
$1,200,000

FEATURED SPACE DATA:

ROOM VOLUME:
49,920 FT3

FLOOR PLAN AREA:
3,120 FT2

SEATING CAPACITY:
100

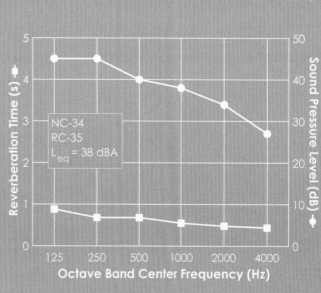

NC-34
RC-35
L_{eq} = 38 dBA

A compact site challenged the goal of peaceful coexistence for a building that improbably combines the performing arts spaces with athletic training rooms and a gymnasium used for NCAA competition. Clever placement of structural isolation joints, attention to structural resonant frequencies, and integration of spring and neoprene isolators help to keep the peace.

Glass walls enclose the theatre, their potentially harsh acoustic effect moderated by a wooden screen that adapts in form as it wraps the room -- emphasizing reflection near the stage and diffusion behind the audience. The result is a vibrant, transparent space that retains acoustic warmth and support for those inside. Interior glazing, located behind the side seating nearest the stage, extends from floor height to the underside of the balcony. By narrowing the front of the room, these surfaces play an important role in delivering acoustic intimacy to the audience. The glazing is canted in section by 1/8" per foot, sufficient to prevent flutter.

Additional shaping is provided at the rear wall and at the glazing which flanks the lobby doors. The rear wall is canted in section, directing energy towards the underside of the balcony, which is in turn shaped to return energy back towards the rear audience members. The glazing flanking the doors is angled in plan, to prevent flutter between this surface and the exterior windows. These shaping elements can be seen in both plan and section details. In both instances, the angled elements are visually masked by the wood grillage – a move which maintains an orthogonal geometry for the eye despite the subtle shaping for the ear.

A forestage reflector shares the simplicity of geometry and materiality as the rest of the theatre. Its flat surface is modestly angled to provide an overhead reflection to the audience seated on the main floor, while giving a nod to the adjacent horizontal metal mesh ceiling elements. An orchestra shell shares the same wood vocabulary as the forestage reflector, wood grillage and seats, providing a cohesive visual warmth that accompanies the acoustic support and envelopment.

The reverberation time data shown were calculated for the occupied space. The background noise level data shown were simulated for the unoccupied space with the HVAC system on.

Music, drama, and dance rehearsal rooms live right below the theatre, each daylit with ample glass that conveys an attitude of extroversion and community. Subtle moves were taken to prevent flutter conditions between nearly-parallel surfaces: Interior glazing assemblies are canted in section, sending energy up towards the ceiling; gentle chevron shaping is incorporated into opaque walls and doors. Despite the glass and the adjacencies, each room is thoroughly isolated from the others, inviting boisterous activity. Drywall barrier ceilings separate the theatre from spaces below, at times taking on an unusual stair-step geometry which maximizes the acoustic volume of the lower space while maintaining a relatively tight floor-to-floor dimension demanded by an overall

Music Rehearsal | Robert Benson Photography

Dance Rehearsal | Robert Benson Photography

building height restriction. An exterior-grade barrier ceiling separates the theatre from the parking garage ramp, located directly below the arbor pit, as can be seen in the building section.

The theatre's Audio and Video systems facilitate a variety of student-run performances and presentations by way of a professional-grade digital mixing console and loudspeaker system, deployed alongside a flexible digital video projection system. A dedicated audio network not only handles the distribution of audio to other rehearsal and support spaces, but also provides operators with the ability to adapt to different productions' needs, whether that involves mixing from a different position in the theatre or operating the systems without technical expertise. While the systems are designed to support practical education by giving students opportunities to work with professional equipment and processes, an interactive control system accessed via touchscreen allows for easy operation of the systems when the technical needs are less complex.

Each of the rehearsal rooms contains a similar system customized to suit the unique functional and architectural needs of each space. For example, the music rehearsal room includes a stereo music

playback system, while the dance rehearsal studio's distributed overhead loudspeaker system provides more uniform coverage of the space while sacrificing the spatial imaging that plays a limited role in dance performance. Video presentation, archival recording, and wireless collaboration devices round out the systems' capabilities for day-to-day educational use. Similar systems for audio and video playback are provided in a variety of classrooms and conference rooms, emphasizing ease-of-use and user interface familiarity in each room.

In addition to the architect and acoustical consultant, the design team included: Theatre Projects Consultants (theatre consultant), Threshold Acoustics LLC (audio systems designer, video systems designer), Horton Lees Brogden Lighting Design (lighting designer), Rist-Frost-Shumway Engineering (mechanical engineer), and Lee Kennedy (general contractor).

Above: Assembly Hall | Robert Benson Photography

THE WINSOR SCHOOL

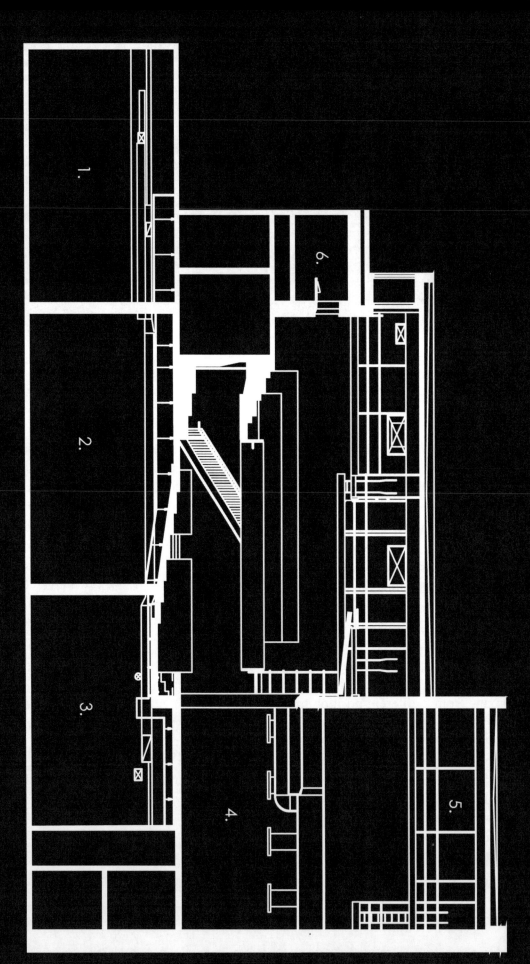

1. Music Rehearsal 2. Dance Rehearsal 3. Drama Rehearsal 4. Stage 5. Loading Gallery 6. Control Booth

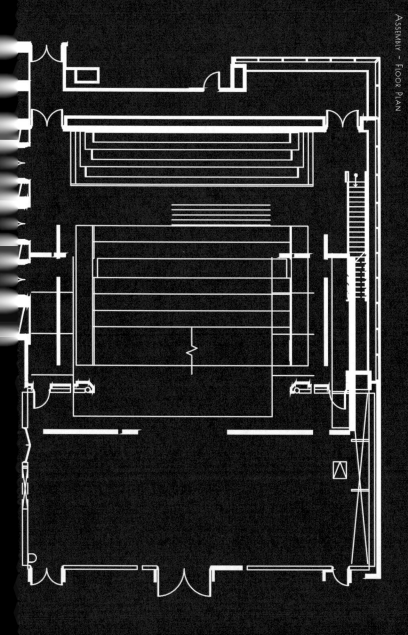

ASSEMBLY – FLOOR PLAN

ACOUSTICAL CONSULTANT:
THRESHOLD ACOUSTICS LLC

ARCHITECT:
WILLIAM RAWN ASSOCIATES

COMPLETION DATE:
2014

LOCATION:
BOSTON, MA I USA

CONSTRUCTION TYPE:
NEW CONSTRUCTION

CONSTRUCTION/RENOVATION COST:
$71,000,000

FEATURED SPACE DATA:

ROOM VOLUME:
175,000 FT3

FLOOR PLAN AREA:
5,000 FT2

SEATING CAPACITY:
515

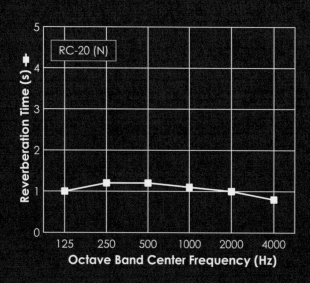

Zumix is a community youth center in the East Boston neighborhood, focused on music. The program acquired an old fire house and sought to make maximal use of its space in order to serve as many people as possible. The resulting design located a multi-purpose performance space on the ground floor, a recording studio and offices on the 2nd floor, and practice rooms, a beat lab, a radio station, and a multi-purpose room on the basement level. With limited budget and space, the priority throughout was to maximize program and minimize cost.

The performance hall occupies the large ground floor space, once used to store fire trucks, and has two large sets of paired doors to the street. The original doors were replaced with new custom-built doors to fit the original arched openings. A flat wooden floor with no fixed furniture was used for the audience area, which allows for great flexibility for the community to use for various gatherings, rehearsals, dances, and other activities. The wooden stage, however, was raised and includes conduit and wiring for power and audio connections. A large retractable curtain covers the upstage wall and provides relief of loudness and improved clarity for louder ensembles. When retracted, the original brick wall is exposed, which is useful for unamplified music. Sound-absorbing wall panels are placed at various locations along the side walls to control flutter echo and other problematic reflections from drums and other percussive sound sources on stage. It is noteworthy that the reverberation times were measured in an unoccupied setting prior to installation of these sound absorbing

features. Apart from these sound-absorbing features, the room's acoustics are largely controlled by the audience condition. The unoccupied room is quite reverberant, but this condition rarely occurs during performances or rehearsals.

The performance hall is equipped with a robust sound system including left and right clusters (JBL, 2-box plus subwoofer each side), for stage monitor floor wedge loudspeakers, and a 16-channel Allen and Heath mixing desk. Wired connections to the recording studio above and the radio station below allow many participants throughout the building to be involved in any performance.

The recording studio includes separate live and control rooms and a vocal booth. These rooms are separated by double stud walls, acoustically rated

doors by IAC, and a custom double-frame window between the control and live rooms. Acoustical treatments include a mix of four-inch and two-inch thick fabric-covered fiberglass panels, diffuser panels by RPG, a corner bass trap in the vocal booth, retractable heavy velour drapery in the live and control rooms, and acoustical ceiling panels. All spaces in the building, including stairs, include wired connections to the recording studio control room so that students can experiment with recording tracks in different acoustical environments. The studio spaces have been particularly well-received, with one user calling it "the best small space that does not sound like a small space that we have ever made recordings in." Basement level music rooms include sound-isolating gypsum board ceilings and a mix of two-inch and four-inch thick fabric-wrapped fiberglass panels surface mounted to walls and ceilings.

ZUMIX received the 2011 National Arts and Humanities Youth Program Award which recognized ZUMIX as one of the top 12 youth arts and humanities after-school programs in the country. The project also achieved LEED Gold certification.

In addition to the architect and acoustical consultant, the design team included: Acentech Incorporated (audio systems designer, video systems designer), Crossfield Engineering (mechanical engineer), and Landmark Structures Corp. (general contractor). The noise rating was calculated for the unoccupied space.

Opposite page: Performance Hall | Erik Jacobs

RADIO STATION | ERIK JACOBS

DRUM ROOM | ERIK JACOBS

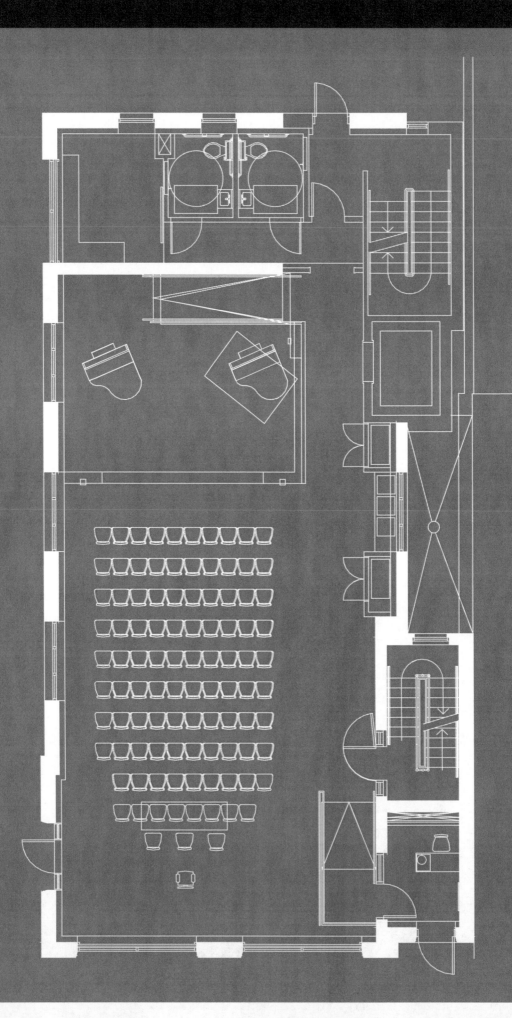

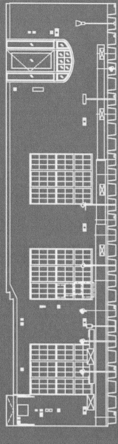

PERFORMANCE HALL – WEST ELEVATION

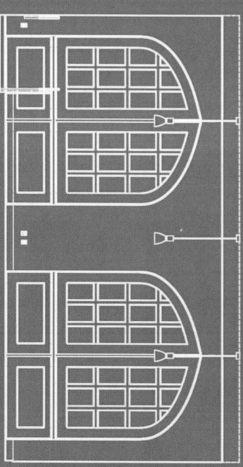

PERFORMANCE HALL – SOUTH ELEVATION

ACOUSTICAL CONSULTANT:
ACENTECH INCORPORATED

ARCHITECT:
UTILE, INC.

COMPLETION DATE:
2010

LOCATION:
EAST BOSTON, MA I USA

CONSTRUCTION TYPE:
RENOVATION

CONSTRUCTION/RENOVATION COST:
$4,600,000

FEATURED SPACE DATA:

ROOM VOLUME:
26,000 FT3

FLOOR PLAN AREA:
1,650 FT2

SEATING CAPACITY:
125

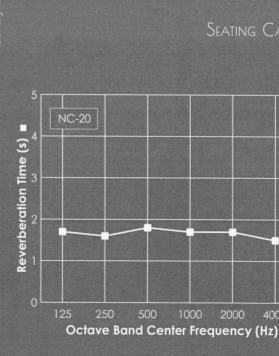

HIGHER EDUCATION FACILITIES:
COMPLETED 2000-2014

The music program at Bethune-Cookman University enjoys a rich tradition immersed in exciting performances and cultural enrichment and has grown into one of the most recognized and most comprehensive music programs in the Southeastern United States. The expanding music program required additional new rooms including the band rehearsal room, percussion room, choral room, instrumental rehearsal room, in addition to smaller ensemble and practice rooms. Bethune Cookman is the home to a variety of classical to contemporary music ensembles including the Marching Wildcats who are dedicated to contributing their artistry and enrichment to the campus and local and regional communities. The marching band known as "The Pride" consisting of 325 members including instrumentalists, a Sophisticated Flag Corp and the

14 Karat Gold Dancers are known beyond the local community through countless nationally televised sporting events and exhibitions.

The pre-renovation Band Room suffered from excessive reverberation from improperly treated walls and a low acoustical ceiling grid system with no sound diffusion. The acoustical volume was increased by opening the ceiling to the deck with suspended field fabricated sound diffusing panels to provide diffuse sound reflections to enhance the communication between the instrumentalists and between the instrumentalists and the music director. The curved ceiling panels that are strategically radiused and angled to direct sound to the musicians allow overhead arrival of diffuse sound to the band members to assist in timing and allow the director

BAND ROOM CEILING | SIEBEIN ASSOCIATES, INC.

BAND ROOM SEATING LAYOUT | SIEBEIN ASSOCIATES, INC.

BAND ROOM | SIEBEIN ASSOCIATES, INC.

BAND ROOM AT AN ANGLE | SIEBEIN ASSOCIATES, INC.

to hear each individual instrument clearly. The emersion of sound then envelops the room together with the interaction of softened sound reflections from the ceiling panels and sound diffusing wall panels. Clarity of sounds are improved by deliberately providing broadband sound absorption through a matrix of sound absorbing panels and field fabricated bass traps which reduce repeated low frequency sound reflections. The balanced acoustical shaping and absorption then begets the collaboration between individual instrumentalists and the band director to become one entity that is the ensemble. The symbiotic relationship between the room and the training musician is thus achieved.

The architect and the college wanted an environment that was visually bright and modern and therefore the underside of the exposed roof deck was covered with sound absorbent material that was white in color. The layout of the band was carefully studied to strategically locate both sound diffusing and sound absorbing panels that allow 'softened' but clear sound reflections so that students can distinguish the sound of his or her own instrument amongst the group, and the instructor is able to listen with clarity to the music. The renovations also included reducing harsh focused sound reflections previously present in the hexagonal Instrumental Room by strategic reshaping of the ceiling plane with angles and sound diffusing panels and sound absorbing ceiling tiles as well as the use of strategic placement of sound absorbent material behind a rich wood grille wall system. The shaped ceiling also allowed increased volume for a

brighter acoustical sound field as well as enhanced arrival of early diffuse sound reflections between the students and the instructor so that each instrument can be distinguished aurally.

The Percussion Room was specifically designed to balance the loudness of drums and other loud instruments and clarity of each beat that is the basis of the marching band. Suffering excessive loudness and fatigue, the percussion students now have their own space that not only allows them to rehearse longer but also time their music in unison and with strength with softened sound reflections from sound diffusing ceiling and wall panels. High performance ceiling tiles with specifically tuned spacing of sound diffusing panels on the ceiling and walls with low frequency absorbing panels and bass traps provide an optimized sound field for the students and instructors alike. The air-conditioning system noise was reduced to provide a quiet environment for all of the music spaces, and sound bleed between rooms was also mitigated through implementing isolated wall assemblies with isolation joints. The reverberation time data shown were calculated for the unoccupied space.

In addition to the architect and acoustical consultant, the design team included SL Construction & Remodeling, Inc. (general contractor). The enhanced environment was successful since both the faculty and the students now take pride in their space that helps fuel their passion for their musical abilities.

PERCUSSION ROOM | SIEBEIN ASSOCIATES, INC.

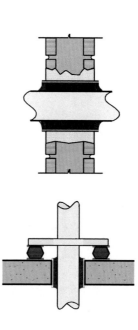

PIPE PENETRATION DETAIL | SIEBEIN ASSOCIATES, INC.

1. BAND REHEARSAL 2. PERCUSSION 3. PRACTICE ROOM 4. MECHANICAL

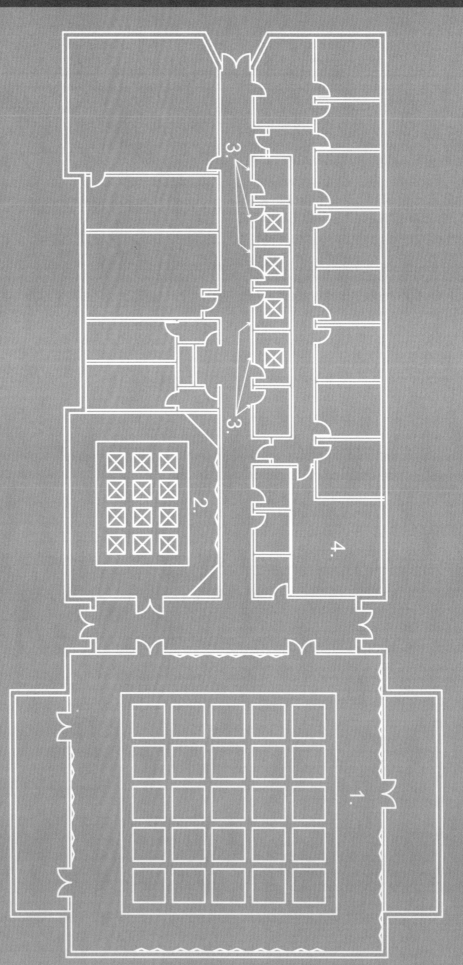

ACOUSTICAL CONSULTANT:
SIEBEIN ASSOCIATES, INC.

ARCHITECT:
KARL THORNE ASSOCIATES, INC.

COMPLETION DATE:
2002

LOCATION:
DAYTONA BEACH, FL | USA

CONSTRUCTION TYPE:
RENOVATION

CONSTRUCTION/RENOVATION COST:
$2,825,000

FEATURED SPACE DATA:

ROOM VOLUME:
93,240 FT3

FLOOR PLAN AREA:
5,160 FT2

SEATING CAPACITY:
150

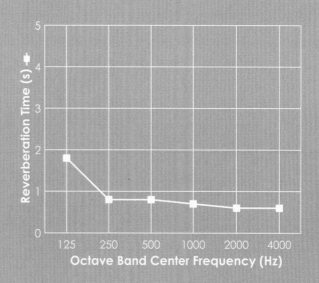

Constructed in 2013, the Casper College Music Building is home to the college's music department. The facility includes the 417-seat Wheeler Concert Hall, ten practice rooms, ten faculty studios, a recording suite, a lecture classroom, administrative spaces, and four large, dedicated rehearsal rooms for percussion, choir, instrumental, and piano.

The lower walls of Wheeler Concert Hall are finished with stacked stone and are angled to prevent flutter echo and provide strong early lateral reflections to the audience. Fabric-wrapped, two-inch thick acoustical panels are applied to the entire lower rear wall. The upper walls consist of horizontal bands of curved, 3/8-inch thick Glass Fiber Reinforced Gypsum (GFRG) panels to diffuse incident sound. Motorized, heavy

drapery, stacked in pockets behind the upper walls, can be deployed around the hall to a total area of 3,200 ft^2 at 100% fullness. The drapery decreases the mid-frequency reverberation time from 1.8 seconds (fully retracted) to 1.2 seconds (fully deployed) to cater the acoustic response of the hall to various performance requirements. The reverberation time data were measured in the unoccupied space.

Six rows of curved, GFRG ceiling clouds are positioned approximately 25-29 feet above stage level and angled to direct sound back down to the platform and audience while also concealing the three technical lighting catwalks. Approximately 45% of the ceiling footprint is open to the deck, which has an average height of 36 feet above platform level.

Supply air is provided to the hall by overhead

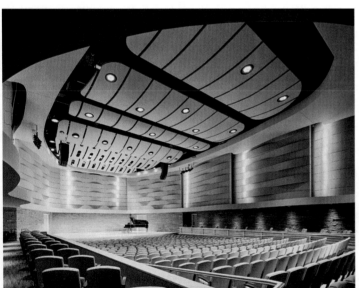

Wheeler Concert Hall | Michael Robinson, c/o MOA

Building Exterior | Michael Robinson, c/o MOA

Wheeler Concert Hall – Drapery Retracted | Michael Robinson, courtesy of (c/o) MOA Architecture (MOA)

ductwork served from a remotely located air-handling unit. Return air is routed under the platform. The background noise level in the unoccupied concert hall was measured to have a rating of RC 21 (HF) with the HVAC system operating. The hall supports amplified music with a high-performance sound reinforcement system that can be controlled from a main control room as well as an alternate in-house mixing position. The system utilizes digital signal transportation between the stage inputs, control booth, and amplifier processing racks and provides a full split of all concert hall inputs to the recording control room.

The instrumental and choral rehearsal rooms feature tall ceilings (23 feet and 24 feet, respectively) and manually-operated drapery to adjust reverberation time. Additional acoustical finishes include fiberglass acoustical ceiling tile, two-inch thick fiberglass wall panels, wall-mounted and lay-in barrel-shaped diffusers. The choral rehearsal room also functions as a small recital hall. The percussion rehearsal room includes curved, reflective clouds suspended from the ceiling and resonant concrete block absorbers with a sound-diffusing face along the upper walls to control low-frequency reverberation. A two-inch expansion joint acoustically isolates the instrumental and percussion rehearsal rooms from the rest of the building. The practice rooms are isolated from one another by double-stud wall construction. To minimize sound flanking between practice rooms, each sits on a resiliently floated floor structure and has a secondary gypsum board lid above the acoustical lay-in finish ceiling.

On the second floor is a recording suite, which includes a control room, studio, and storage space. The control room equipment is capable of recording inputs from the Wheeler Concert Hall, recording studio, and music faculty studios. Each faculty studio has tie lines for microphones as well as connections to the recording studio's headphone monitoring system. The recording studio control room has several dedicated displays including a large display that can switch between a dedicated point-of-view camera in the concert hall and a camera above the control room's main console, which allows recording instructors to present real-time recording console adjustments to the entire control room for instructional purposes.

The Casper College Music Building is primarily used by the music department's students and faculty for instruction, practice, and performance. It also served as the temporary home to the Wyoming Symphony Orchestra for one season during a renovation of their permanent space.

In addition to the architect and acoustical consultant, the design team included: Theatre Collaborative Consultants (theatre consultant), D.L. Adams Associates (audio systems designer, video systems designer), Engineering Design Associates (mechanical engineer), Martin/Martin Wyoming (structural engineer), and West Plains Engineering (electrical engineer).

PERCUSSION REHEARSAL | MICHAEL ROBINSON, C/O MOA

RECORDING CONTROL ROOM AND STUDIO | MOA

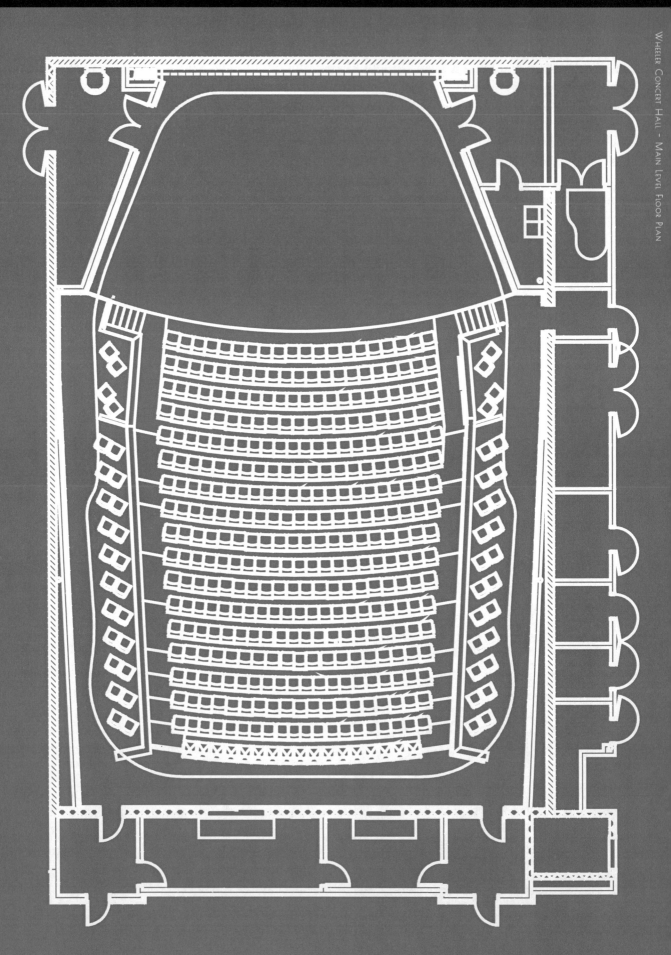

WHEELER CONCERT HALL - MAIN LEVEL FLOOR PLAN

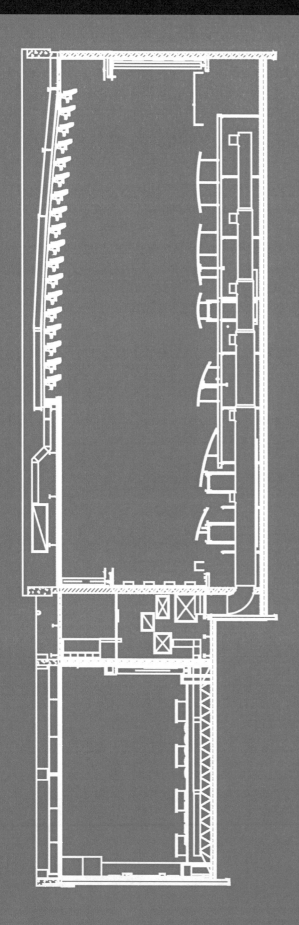

Wheeler Concert Hall – Longitudinal Section

ACOUSTICAL CONSULTANT:
D.L. ADAMS ASSOCIATES

ARCHITECT:
MOA ARCHITECTURE WITH
HMS ARCHITECTS

COMPLETION DATE:
2013

LOCATION:
CASPER, WY I USA

CONSTRUCTION TYPE:
NEW CONSTRUCTION

CONSTRUCTION/RENOVATION COST:
$14,000,000

FEATURED SPACE DATA:

ROOM VOLUME:
190,000 FT3

FLOOR PLAN AREA:
5,200 FT2

SEATING CAPACITY:
417

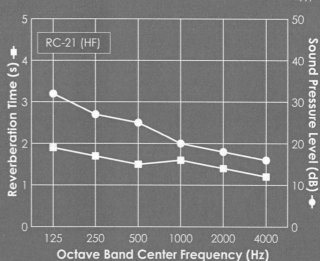

The Center for Visual and Performing Arts (CVPA) at Earlham College is the home of the College's Music and Art Departments. It hosts the Lingle Recital Hall, a flexible performance and rehearsal hall, as well as dedicated rehearsal spaces for the College's wide and unique array of musical ensembles, and a full suite of teaching studios and practice rooms. Working spaces for the Art department host drawing, painting, weaving, photography, and metalworking spaces. A highly configurable black box studio theater supports a broad spectrum of theatrical uses.

The Lingle Recital Hall occupies a rectangular footprint in the center of the building, constructed of heavy masonry for acoustic isolation from nearby acoustically sensitive spaces as well as for a bass-supportive room acoustic signature. Scored and split-face block further shapes the room acoustic signature by providing fine-scale diffusion. Within the perimeter walls, an open wooden basket weave treatment rings the lower occupied region of the room to provide acoustic diffusion. Around the performance platform, the basket weave stands freely inboard of the perimeter wall, with acoustically absorptive drapery available to draw behind to control loudness for large ensembles.

Motorized retractable acoustic banners can be deployed for reverberation and loudness control of large and amplified ensembles. Performance audio and projection systems further broaden and enhance the array of performances that can be supported. A retractable audience seating system can be removed

from the space to bring the room to a fully flat floor condition, allowing a large orchestra to rehearse comfortably. An adjacent recording room, with a direct view to the performance platform, has extensive connectivity into the recital hall (as well as the building's studio theater) for high-quality recording.

The nearby studio theater features the ability to host an extremely wide range of performance configurations, with fully configurable seating platforms as well as a tension grid system that allows lighting, audio, and video systems to be reimagined and optimized for any event. A split-face masonry box again performs multiple duties both for acoustic isolation and interior diffusion. Tectum panels of varying thickness provide reverberation control and prevent flutter in the occupied lower stratum of the room.

The comparatively small footprint of the building placed jazz and percussion rehearsal rooms directly adjacent to each other and across a corridor from the recital hall. Each of these rooms is constructed on its own independent foundation, with full structural separation from the remainder of the building, allowing ensembles in both spaces to rehearse at full tilt without disturbing an intimate recital across the hall. Structural isolation bearings at the roof level complete the separation.

A dedicated rehearsal suite for the department's Javanese gamelan program gives both dedicated rehearsal and storage space, with independent humidity control to preserve the well-being of its unique and prized instruments. A tall volume with a low height reflector array accommodates the energy output of the full ensemble but preserves good cross-ensemble hearing conditions.

Voice and piano teaching studios, as well as a full suite of instrumental and dedicated percussion practice rooms and faculty offices, support the pedagogical needs of the music department. Isolated gypsum board boxes with wall-supported cap ceilings in these areas allow large trunk-line ductwork serving the major performance and rehearsal spaces to pass above without risk of being affected by mechanical systems' noise and vibration.

A slab isolation joint, isolated stud walls, and a low-profile acoustic barrier ceiling enclosing a metalworking studio protect the nearby recital hall from the hammer blows of artistic vision.

In addition to the architect and acoustical consultant, the design team included: Schuler Shook (theatre consultant, lighting designer), Threshold Acoustics LLC (audio systems designer, video systems designer), dbHMS (mechanical engineer), Shiel Sexton (general contractor), and Halvorson and Partners (structural engineer). The reverberation time data shown were measured in the unoccupied space. The background noise level data shown were calculated for the unoccupied space.

Opposite page: Recital Hall - Interior | Steve Maylone

Percussion Rehearsal | Steve Maylone

Diffusive Grillage | Steve Maylone

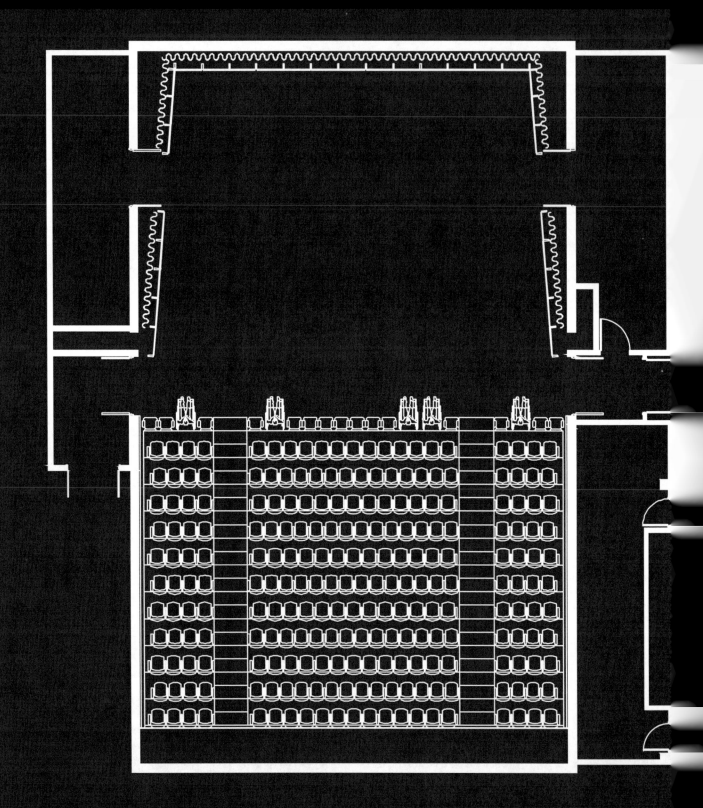

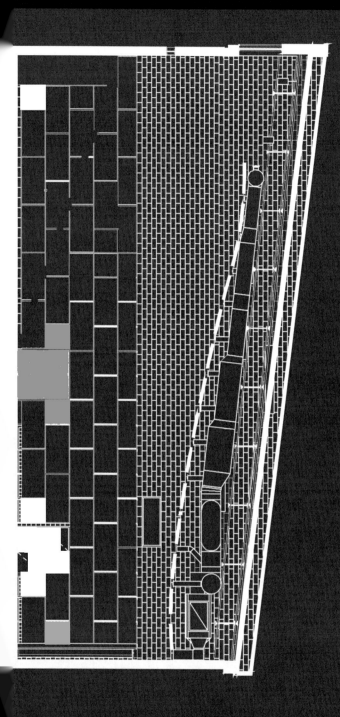

RECITAL HALL – SECTION

ACOUSTICAL CONSULTANT:
THRESHOLD ACOUSTICS LLC

ARCHITECT:
BOORA ARCHITECTS

COMPLETION DATE:
2014

LOCATION:
RICHMOND, IN | USA

CONSTRUCTION TYPE:
NEW CONSTRUCTION

CONSTRUCTION/RENOVATION COST:
$17,600,000

FEATURED SPACE DATA:

ROOM VOLUME:
138,000 FT3

FLOOR PLAN AREA:
4,000 FT2

SEATING CAPACITY:
260

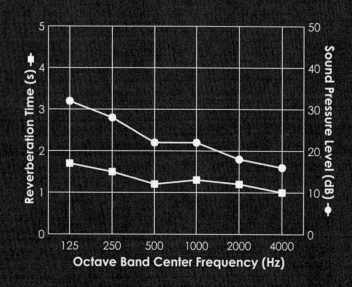

The Performing Arts Building on the Ybor City Campus at Hillsborough Community College was a comprehensive addition and renovation project that included major acoustical improvements to the existing building as well as the acoustical design for new music rehearsal spaces. Extensive acoustical enhancements were integrated within the building that consisted of optimized sound field design in performing, rehearsal, and practice spaces and mechanical system noise control. Sound isolation upgrades were also an integral part of the design in spaces for musical and theatrical rehearsals and performances including the new band room, practice rooms, faculty teaching studios, and the complete renovation of the existing theater, black box theater, and classrooms.

Band rooms are usually too loud when they are fully occupied. They also generally suffer from a lack of sound diffusing elements on the walls and ceiling of the room to allow good hearing among the students while they practice. The existing band room suffered acoustically from low ceiling height, lack of acoustical shaping and finishes that resulted in a loss of clarity from the cacophony of many instruments playing in a small space with harsh and uneven reflections. Sound transmission was also a challenge between adjacent spaces both vertically and horizontally where the students in various practice rooms could hear each other playing at disturbing levels through the walls and ceilings. The section drawings show how classroom spaces for music education and offices are embedded within the theater and the black box theater. The existing theater and black box theater,

and other music rehearsal spaces also had similar acoustical challenges inherent with the original design and construction of the multi-story building. Sounds from the piano lab were clearly heard in the black box theater below, and the noise from the rooftop air-handling units propagated into the stage below.

The band room design included removing the existing ceiling and extending the height to 25 ft, a huge improvement over the existing space. The additional volume allowed sounds to properly dissipate in the space and the careful shaping and angling of the overhead panels allowed the students to clearly hear each other and the instructor. A unique aspect of the design included a huge window on the plan north wall to allow plenty of natural light in the space. The window was designed to be angled outward towards the top, so as to provide diffusion, despite being a largely reflective surface. Gentle curved reflectors were used on the lower portions of the side walls that direct sound from the performers back into the space, allowing the students and instructors to clearly hear each other and providing uniform sound decay in the room.

The ensemble rooms will be used for music rehearsals by smaller groups of students. Similar to the band room, sound diffusing panels on the walls and ceiling were incorporated for students and the instructor to hear each other. Strategic areas of acoustic ceiling and acoustic wall panels were selected to limit reverberation and sound build-up. Since the ensemble rooms would also be used for choral rehearsal, which prefer to be slightly more bright and 'live' than ensemble rooms used for choir, several schemes for a band ensemble and a choral or orchestral ensemble were presented to the architect.

Some of the acoustical challenges addressed through strategic planning included relocating music rehearsal spaces away from the theater and black box theater for improved sound isolation. Space shaping and finishes, such as convex-curved sound diffusing ceiling and wall panels to provide sound reflections for clarity and texture of music were also integrated into the design of each of the spaces. Strategically locating sound absorbent finishes and an enlarged acoustical

volume resulted in the provision of a desired sound field for music instruction and rehearsals.

Additionally, the band room, studio and practice room additions provided acoustically improved spaces for music instruction, and created flexibility in the faculty's music instruction as well as increased opportunities for the expansion of their performing arts program.

In addition to the architect and acoustical consultant, the design team included Robert Lorelli Associates, Inc. (theatre consultant). The reverberation time data shown were measured in the unoccupied space. The noise ratings shown were calculated for the unoccupied space.

Opposite page: Band Room | Siebein Associates, Inc.

THEATER – STAGE | SIEBEIN ASSOCIATES, INC.

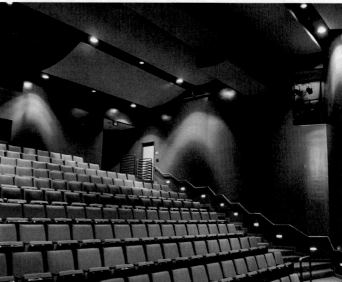

THEATER | WILLIAMSON DACAR ASSOCIATES

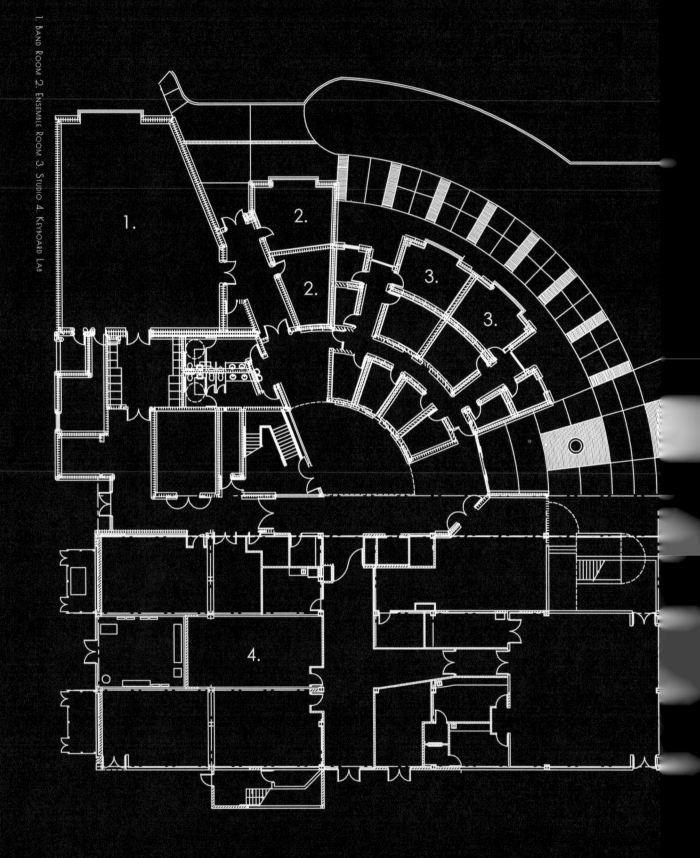

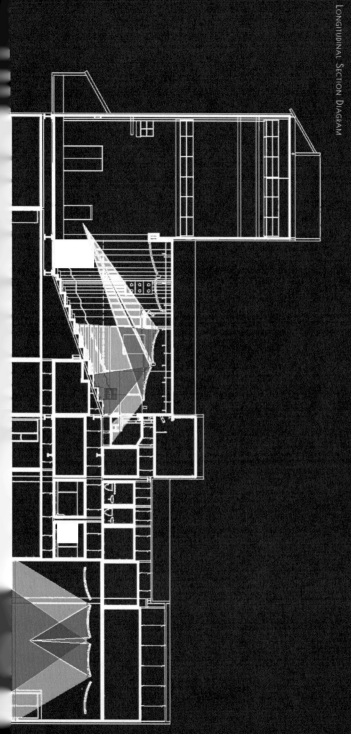

LONGITUDINAL SECTION DIAGRAM

ACOUSTICAL CONSULTANT:
SIEBEIN ASSOCIATES, INC.

ARCHITECT:
WILLIAMSON DACAR ASSOCIATES

COMPLETION DATE:
2008

LOCATION:
TAMPA, FL | USA

CONSTRUCTION TYPE:
NEW CONSTRUCTION, RENOVATION,
EXPANSION

CONSTRUCTION/RENOVATION COST:
$7,400,000

FEATURED SPACE DATA:

ROOM VOLUME:
65,286 FT3

FLOOR PLAN AREA:
2,124 FT2

SEATING CAPACITY:
246

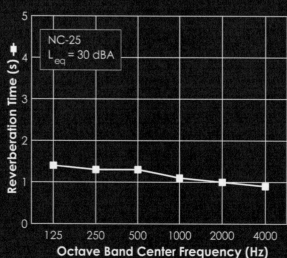

NC-25
L_{eq} = 30 dBA

Poindexter Hall at the Mississippi University for Women (MUW) was completed in 1905, nicknamed "The Temple of Music" and eventually named after Miss Weenona Poindexter, who started a conservatory-style music program on campus in 1894. The building was marked by the presence of President Taft in 1909 on his victory tour where he "stood with II&C president Henry Whitfield to show his support for the college's mission. "A girl has the right to demand such training that she can win her own way to independence, thereby making marriage not a necessity, but a choice," the President said." [1] The building almost perished during a massive fire of the neighboring four-story Shattuck Hall on July 15, 1953; however, Miss Poindexter (retired, in her 80's) entered Poindexter Hall to ensure its protection due to it being occupied.

The renovation was completed in 2012, and the hall hosts a variety of recital and concert programming with lectures, but the predominant programming consists of piano and vocal recitals and choir concerts, which is the mainstay of the MUW program. The facility consists of three floors on the front of the building and four floors at the back, housing twelve practice rooms, nine faculty studios, a choir room, an electronic piano lab, an instrumental music lab/rehearsal/recording room, a grand piano studio, full backstage facilities, and a "B-grade" recording studio wired directly to the recital hall and several practice studios as well as a main tracking room.

Interesting features include the 2nd floor hallway being open to the hall, facilitating a modest double slope decay due to the geometric volume and finishes

2ND FLOOR HALL-BALCONY | CHRIS JENKINS

CHOIR ROOM | CHRIS JENKINS

EARLY FIELD TESTING | DAVID WOOLWORTH

of the hall. Also, the entire downstairs of the hall is surrounded by french doors, which allow for daylighting and a visual connection to the outside (and an audible connection to both the BNSF and the Columbus & Greenville railways, which run on the edge of campus, about 550ft away).

Some of the challenges in the Poindexter project include getting good seals on doors to studios and practice rooms with irregular historical floors and modifying historical doors to improve sound transmission performance while maintaining appearance for the Department of Archives and History. The oval shaped recital hall was one of the largest acoustical challenges, treated with "sound soak" panels in the 1970's, newly treated with deep diffusion panels covered with fabric to blend with the architecture and provide modest additional sound absorption. Point source loudspeakers are embedded in between the diffusion panels behind the acoustically transparent fabric, and are used for reinforcement of speaking voice or pre-recorded audio. The reverberation time data were measured in the unoccupied space. The background noise level data were measured in the unoccupied space with the HVAC systems on, water fountains off, and stage lighting off.

In addition to the architect and acoustical consultant, the design team included: Roland, Woolworth & Associates, LLC (audio systems designer, video systems designer, lighting designer), Pryor Morrow Architects and Engineers (mechanical engineer), and West Brothers Construction (general contractor). The project received an Award of Recognition for Renovation for Historical Preservation by the International Interior Design Association, and the Silver Award for Renovation for Historical Preservation by the American Society for Interior Designers.

[1] Sobolewski, Rich. "The 'Temple of Music' Reopens". *Visions, Mississippi University for Women.* www.muw.edu/visions/features/120-the-temple-of-music-reopens. Accessed 20 December 2019.

Opposite page: Recital Hall | Chris Jenkins

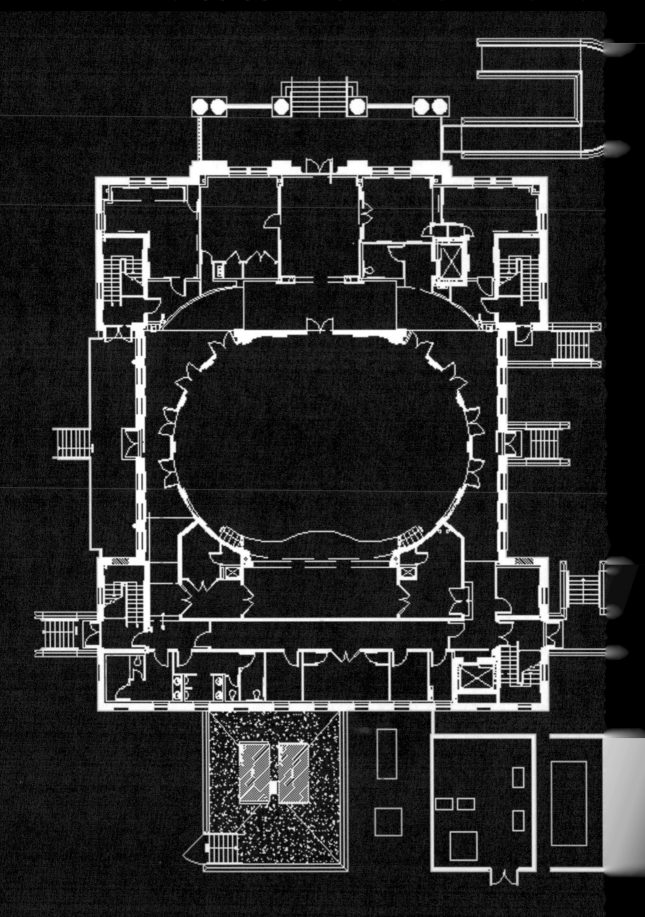

ACOUSTICAL CONSULTANT:
ROLAND, WOOLWORTH & ASSOCIATES, LLC

ARCHITECT:
PRYOR MORROW ARCHITECTS AND ENGINEERS

COMPLETION DATE:
2012

LOCATION:
COLUMBUS, MS I USA

CONSTRUCTION TYPE:
RENOVATION

CONSTRUCTION/RENOVATION COST:
$9,500,000

FEATURED SPACE DATA:

ROOM VOLUME:
87,464 FT3

FLOOR PLAN AREA:
3,364 FT2

SEATING CAPACITY:
152

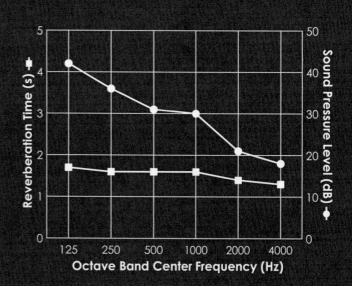

The president of North Central College, located in the heart of Naperville which is Illinois's fifth largest municipality, sought to build a concert hall in spite of not having an orchestra as a prime user. This was an example of "If you build it . . they will come." [1]

Dr. Myron Wentz, a past graduate of the school, was the primary donor and his name graces the hall. Developed with the goal of excellent acoustic quality, a design was created which favored *clarity* plus *reverberance*, both *running liveliness* and *terminal reverberance*.

The stage area and the associated acoustic volume was established to accommodate a moderately large orchestra of approximately 85 musicians and a 100-voice chorus. In contrast to typical thinking of developing the acoustic volume on a "volume per patron" basis, it is the opinion of the acoustician that the acoustic volume of a concert hall should be commensurate with the size of the orchestra planned for accommodation. The acoustic volume of the Wentz is approximately 500,000 cubic feet.

To achieve significant *running liveliness* and the ability to hear the *reverberance*, as the music is ongoing, not simply when the music stops, room shaping methods were employed which resulted in high Early Decay Time (EDT) values. A room of moderate width was established, overall room shape was made slightly reverse fan, the rear sections of the hall were boldly reverse-fan shaped at 16-degrees, and other room shaping methods were employed. The intent was to create strong sound reflections in

the 100- to 300-millisecond range. The resulting mid-frequency EDT values were measured and averaged to be 2.6 seconds, unoccupied, with all sound absorbing curtains and banners retracted. The benefits of high EDT values are consistent with the findings of Leo Beranek's significant publication [2] showing excellent correlation between EDT values and subjective preference of 50 concert halls, but note the Wentz was intentionally designed for high EDT prior to the release of Beranek's publication.

Local building height restrictions limited the ceiling height within the hall but terminal *reverberance* was optimized using coupled rooms. Reverberation chambers were developed in locations flanking the stage and a carefully calculated common area was determined using coupled room differential equations. The President liked to say that each chamber is the volume of a typical Naperville home. More specifically, the two acoustic chambers comprised approximately 30% of the total room volume.

The reverberation chambers create a coupled room effect which results in a beneficial dual-slope sound decay. Recall that the dynamic range within (quiet) concert halls far exceeds 60-decibels when energized with loud music. When music stops following a *fortissimo* passage in the Wentz, the rate of decay for sound is much less following the "knee" of the double slope, which occurs at approximately 0.75 seconds in the decay process. As such, the *loudness* of the *reverberance* between -60dB and -80 dB during terminal reverberation remains much more audible versus the decay process within a single chamber concert hall of equal volume.

Clarity of sound within the hall is considered excellent.

Sound diffusion was integrated into the design at locations where specular reflections were not created. RPG *Flutterfree* material was bonded to the side walls on wood slats held away from the base wall by 0", 1-1/2", and 3" with the intent of extending the bandwidth of the system's sound diffusive properties. Special care was taken to contiguously and firmly bond this diffusion system to the grouted CMU wall behind, thereby maintaining low-pitched sound reflection to create a *warmth* of sound.

The result is a lushly reverberant space for the enjoyment of music. An observed characteristic of the hall is that solo performers can work the hall and maximize the *reverberance* by orienting their instrument toward the openings of the reverberation chambers, which flank the stage.

Considerable flexibility was integrated into the hall through the integration of retractable velour curtains and banners. Double-layer 32-ounce velour curtains and banners were integrated into the design to offer significant reduction of EDT and T60 values. Abundant soft goods were needed to tame the very sound reflective reverberation chambers. The retractable absorption is motorized and computer controlled. Available settings were established based on the intended use of the hall, including settings for rehearsal (unoccupied hall) activities.

The background noise level is approximately RC-15, depending on location. The hall is often used for recording purposes.

The Wentz has hosted the Chicago Symphony Orchestra, Milwaukee Symphony Orchestra, Chicago, Sinfonietta, DuPage County Symphony Orchestra and numerous other professional music ensembles.

In addition to the architect and acoustical consultant, the design team included Schuler Shook (theatre consultant).

[1] *Field of Dreams*. Directed by Phil Alden Robinson, Universal, 1989.

[2] Beranek, Leo L. "Concert Hall Acoustics-2008." *Journal of the Audio Engineering Society*, vol. 56, no. 7/8, 2008, pp. 532-544.

Opposite page: View from House Left | R. Talaske

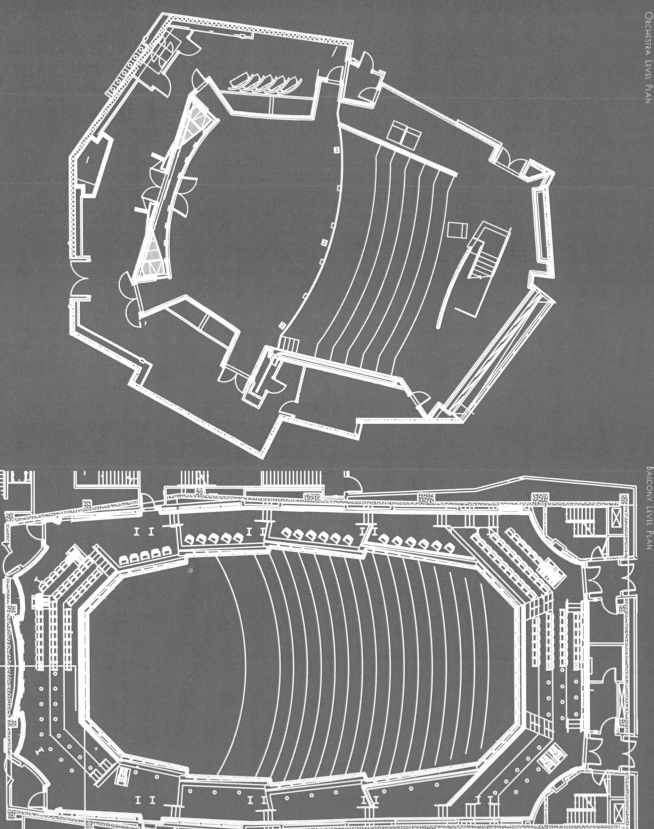

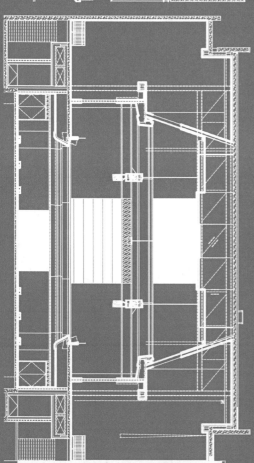

LONGITUDINAL SECTION

CROSS SECTION

LONGITUDINAL SECTION

ACOUSTICAL CONSULTANT:
TALASKE I sound thinking

ARCHITECT:
LOEBL SCHLOSSMAN & HACKL

COMPLETION DATE:
2008

LOCATION:
NAPERVILLE, IL I USA

CONSTRUCTION TYPE:
NEW CONSTRUCTION

CONSTRUCTION/RENOVATION COST:
$29,000,000

FEATURED SPACE DATA:

ROOM VOLUME:
500,000 FT3

FLOOR PLAN AREA:
13,000 FT2

SEATING CAPACITY:
617

Maier Hall is a 65,000 square-foot, three-story multidisciplinary center for arts, humanities and instructional support programs located at Peninsula College in Port Angeles, Washington. The building houses a 135-seat recital hall, a rehearsal room and music practice rooms. Maier Hall was designed to surpass the Architecture 2030 Building Challenge for sustainability, reduced energy use, and fossil fuel dependence, and was awarded LEED Gold rating.

At 60 feet long, 45 feet wide, and 28 feet tall, the multidisciplinary recital hall provides an intimate theater space designed to accommodate music rehearsal and performance, lecture, and film. The side walls and the ceiling are designed of segmented bands shaped to provide reflections to specific seating areas, collectively providing full coverage. The ceiling portion of the band consists of two triangles in a kite-like configuration in a changing pattern and angled to direct reflections down. The side-wall bands and the ceiling are built forward of the hall's textured concrete shell. Velour curtains can be deployed along the side walls between the segmented bands and the concrete shell, and then stow away behind the stage's side walls. The diffusive rear wall is designed with a flutter-free, saw-toothed lower section and textured concrete upper section. Vertically retracting curtains can be deployed along the upper section of the rear wall as part of the hall's variable acoustics scheme. The motorized overhead reflectors above the stage were adjusted in place during the hall's fine-tuning period. The textured concrete walls are painted with heavy bridging paint, and the rear wall flutter-free

REAR WALL WITH BUSTER | STANTEC CONSULTING SERVICES

STAGE OVERHEAD REFLECTORS | STANTEC CONSULTING SERVICES

SIDE AND REAR WALLS | STANTEC CONSULTING SERVICES

wood casing, the side wall reflectors, and the ceiling reflectors were randomly stiffened to vary their response. The motorized curtains on both side walls and the rear wall may be controlled independently of each other, allowing for greater customization of room response. A simple control panel at the lectern allows users to control the curtain settings for more flexible room configurations.

The recital hall is located directly underneath classrooms, and the acoustical separation was focused on the classroom side to maximize the recital hall height and volume. The classrooms were designed with double stud walls floating on an isolated concrete floor.

The reverberation time measured at 500 Hz in the empty hall ranged between 1.27 and 1.81 seconds with curtains fully deployed and curtains stowed away, respectively. The reverberation time with the rear curtain only was 1.61 seconds, with one side wall curtain only was 1.65 seconds and with both side wall curtains only was 1.45 seconds. The noise rating shown was calculated for the unoccupied space.

In addition to the architect and acoustical consultant, the design team included: Auerbach Pollock Friedlander (theatre consultant, audio systems designer, video systems designer), Stantec Consulting Services (lighting designer), WSP Flack + Kurtz (mechanical engineer), Howard S. Wright (general contractor), Magnusson Klemencic Associates (structural engineer), KPFF Consulting Engineering (civil engineer), O-Brien & Co (sustainable building strategies), Westech Co. (environmental consultant), and Nakano Associates (landscape architect).

Opposite page: Building Exterior | Schacht Aslani Architects

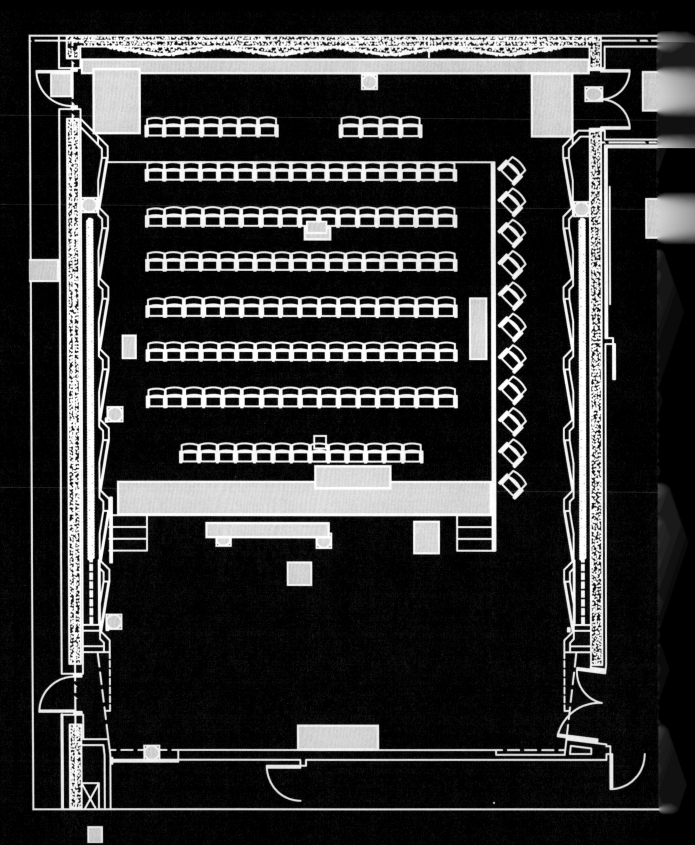

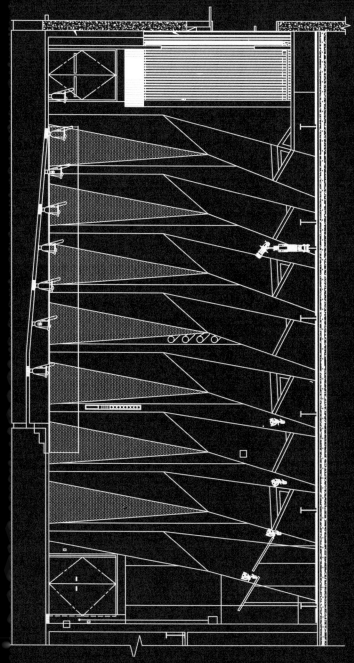

Side Wall Elevation

ACOUSTICAL CONSULTANT:
STANTEC CONSULTING SERVICES

ARCHITECT:
SCHACHT ASLANI ARCHITECTS

COMPLETION DATE:
2012

LOCATION:
PORT ANGELES, WA I USA

CONSTRUCTION TYPE:
NEW CONSTRUCTION

CONSTRUCTION/RENOVATION COST:
$36,000,000

FEATURED SPACE DATA:

ROOM VOLUME:
75,600 FT3

FLOOR PLAN AREA:
2,700 FT2

SEATING CAPACITY:
135

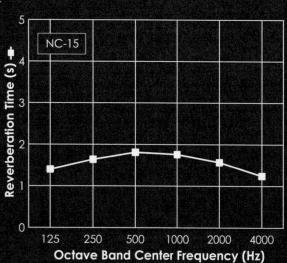

eed College is a private, liberal arts college founded in 1908 outside Portland, Oregon. To better support its goal of broad liberal arts education, Reed College commissioned an interdisciplinary building to house the music, dance, and theater programs in a central location. The Reed College Performing Arts Building is a three-story facility designed to encourage collaboration and innovation between the programs and contains a variety of spaces to suit the needs of musicians, actors, and dancers. A central atrium connects a 200-seat studio theatre, a 100-seat black box theatre, a 120-seat music/choral rehearsal room, and a 100-seat multipurpose performance space. The building also contains a large dance studio, costume design studios, shared classrooms, offices, and a suite of practice rooms for musicians. There are fourteen practice rooms of various sizes, including a group practice room, a dedicated percussion practice room, and individual rooms.

Before the construction of the performing arts building, the choir would hold rehearsals in Reed College's chapel, which has a measured reverberation time of 1.5 seconds. The new music rehearsal room was meant to be a dedicated space for musicians and replaced the chapel as a rehearsal space for medium-sized ensembles.

Located on the third floor, the music rehearsal room is flooded by daylight from large windows on the back wall of the room. Permanent tiers of lightly upholstered, fixed seats (120 in all) face a presentation area, large enough for a small ensemble, piano, or musical director to conduct rehearsals from. When

not in use as a vocal or instrumental rehearsal room, the room is used as a lecture hall.

Balancing the acoustical needs of the room's uses required a design that honored the choir director's wish for a very lively rehearsal space with the need for good speech intelligibility. As the design team planned the space, they moved away from the use of variable acoustic treatments in favor of a fixed, lively environment with a reverberation time of 1.25 seconds. The reverberation time data shown were calculated for the unoccupied space. To offset the effects of the extended room response, large amounts of diffusive surfaces were used to create the ideal signal-to-noise ratio for speech without lowering the reverberation time desired by the musicians. There are very few elements within the space that

cause strong reflections that would interfere with the speech aspect of the room. The noise rating shown was calculated for the unoccupied space.

Both side walls and much of the front are lined with wooden quadratic residue diffusers. Although these surfaces are mildly absorptive, they preserve and diffuse most of the acoustic energy that strikes them. The elevated ceiling consists of curved overhead reflectors spaced away from a hard GWB ceiling to create diffusion and add to the complexity of the reflections from overhead. Interspersed between the reflectors are small areas of absorptive wall panels. The rear wall alternates between solid wood and gypsum board in the areas between the floor-to-ceiling windows. The front wall is primarily quadratic residue diffusers except for a white board and wooden score storage drawers.

In addition to the architect and acoustical consultant, the design team included: Fisher Dachs Associates (theatre consultant), Listen Acoustics (audio systems designer, video systems designer), Luma Lighting Design (lighting designer), PAE Consulting Engineers (mechanical engineer), Hoffman Construction (general contractor), Mayer/Reed (landscape architect & graphic designer), KPFF Consulting Engineers (structural engineer), and Harper Houf Peterson Righellis Inc. (civil engineer).

Opposite page: Rehearsal Room - Front Wall | Stantec Consulting Services

REHEARSAL ROOM - SIDE WALL DIFFUSION | STANTEC CONSULTING SERVICES

REHEARSAL ROOM - CEILING | STANTEC CONSULTING SERVICES

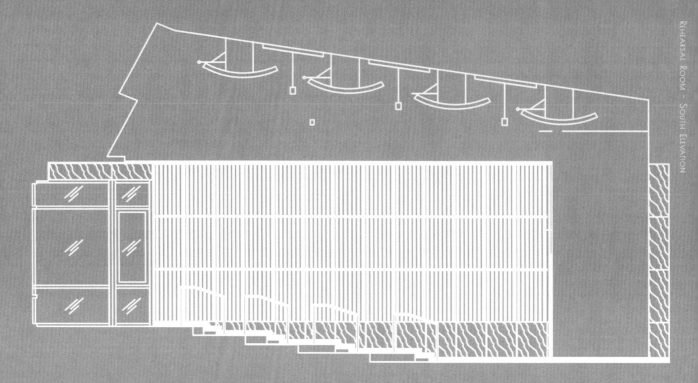

REHEARSAL ROOM – SOUTH ELEVATION

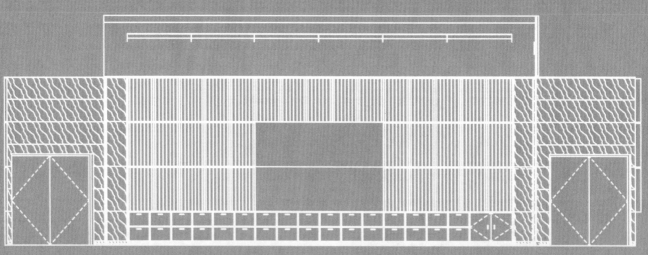

REHEARSAL ROOM – WEST ELEVATION

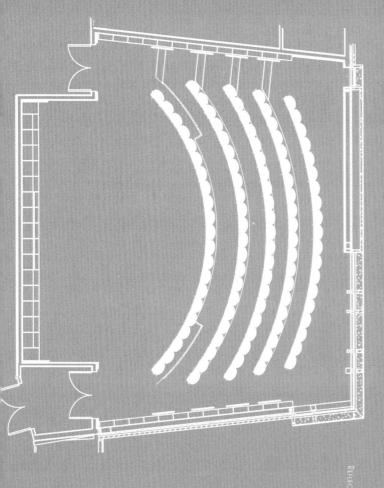

Rehearsal Room - Floor Plan

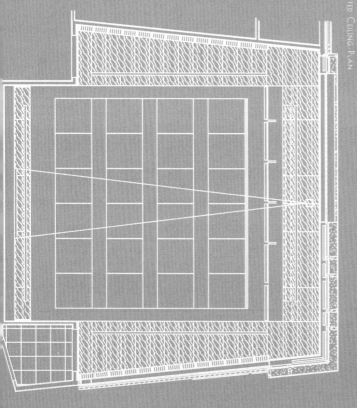

Reflected Ceiling Plan

ACOUSTICAL CONSULTANT:
STANTEC CONSULTING SERVICES

ARCHITECT:
OPSIS ARCHITECTURE

COMPLETION DATE:
2013

LOCATION:
PORTLAND, OR I USA

CONSTRUCTION TYPE:
NEW CONSTRUCTION

CONSTRUCTION/RENOVATION COST:
$28,000,000

FEATURED SPACE DATA:

ROOM VOLUME:
43,000 FT3

FLOOR PLAN AREA:
1,950 FT2

SEATING CAPACITY:
120

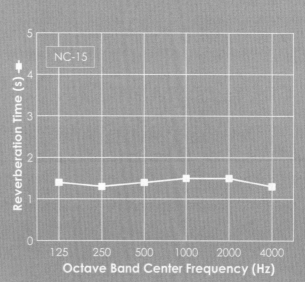

Curtis R. Priem Experimental Media and Performing Arts Center (EMPAC) at Rensselaer Polytechnic Institute (RPI) is a signature work of architecture that brings together four main venues as well as artist-in-residence studios, audiovisual production and post-production suites, student facilities and lab spaces for cross-disciplinary collaboration under one roof. Spaces can be used independently and simultaneously for events, performances, hosting student gatherings, conducting research, and offering other opportunities for interaction and exchange between artists and researchers in science and technology. To accommodate the range of uses RPI required of these spaces, all venues required extensive isolation, variable acoustics, acoustic finishes and quiet mechanical systems.

The 1200-seat concert hall is designed to be a first-rate venue for orchestras and to be equally capable of accommodating presentations with surround sound compositions and video projection. To achieve the variable acoustics required of the space, 13 sets of acoustical banners (34 in all) – five sets along the sides, three sets at the rear wall, and five sets at the upstage wall – allow the concert hall's reverberation time to vary from less than 2.0 seconds with upstage banners deployed and all other banners deployed to ceiling height to 2.7 seconds with all banners fully stored. The reverberation time data were measured in the unoccupied space with the upstage banners fully deployed and all other banners deployed to the fabric ceiling height. Considerable materials research and development went into the design of the ceiling's fabric material, a custom-woven Nomex canvas,

Concert Hall Interior | Kirkegaard

Concert Hall Ceiling | Kirkegaard

Theater Interior | Kirkegaard

to obtain the desired balance of sound reflection, absorption, and transmission. The concert hall sits on a plinth that includes a supply air plenum, used for displacement ventilation, and the dedicated mechanical equipment room, which is structurally isolated from adjacent spaces. The background noise level data were measured in the unoccupied space.

In addition to the concert hall, EMPAC has two flat-floor, experimental theaters. A "multi-modal" environment, Studio 1 - Goodman is an exceptionally versatile space for the integration of technology with human expression and perception. The bespoke two-foot by two-foot acoustic wall panels offer a novel approach to the acoustics of this innovative and immersive sound environment. Behind the panels are tuned bass absorbers to flatten out the room response. Studio 2 has the same concept with a lower ceiling, smaller footprint, and white diffusive pannels. Both studios feature full box-in-box construction, quiet air delivery, and highly networked environments – rewarding audiences and supporting experimentation by artists and researchers alike.

With a 40-foot by 80-foot stage and a 70-foot flytower, the 400-seat Theater provides a facility for experimental artists. It can be used with or without its orchestra pit, and movable seating at parterre level, along the sides, allows artists to configure the theater as a proscenium space or to extend the playing area along the sides of the audience. The framing of the side galleries accommodates the attachment of projection screens and loudspeakers, allowing the audience to be immersed in virtual environments.

In addition to the architect and acoustical consultant, the design team included: Fisher Dachs Associates Inc. (theatre consultant), Kirkegaard (audio systems designer, video systems design), Office for Visual Interaction Inc. (lighting designer), and Davis Brody Bond, LLP (architect of record).

Opposite page: Studio 1 - Goodman Interior | ESTO

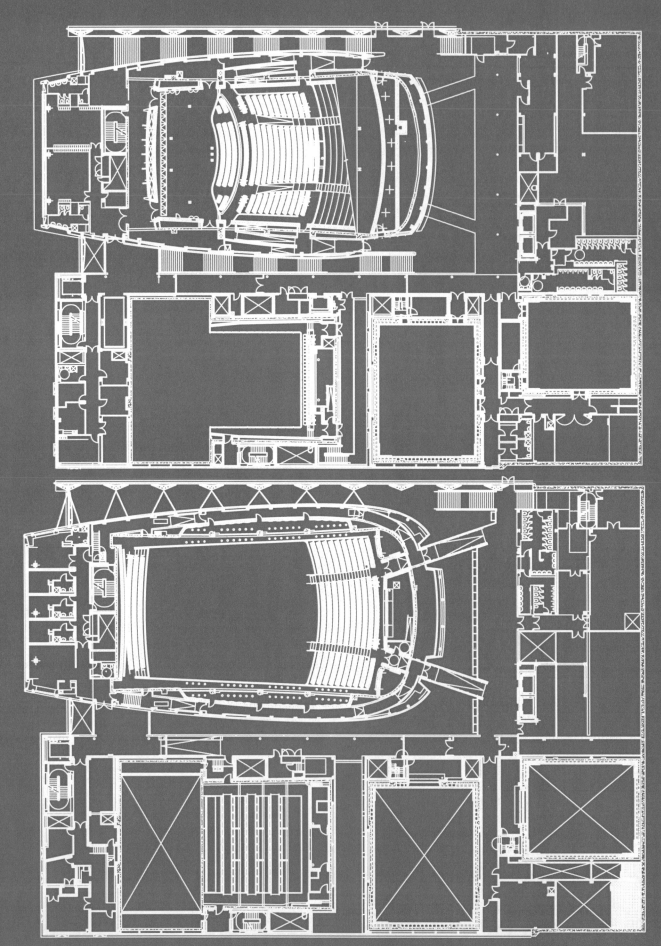

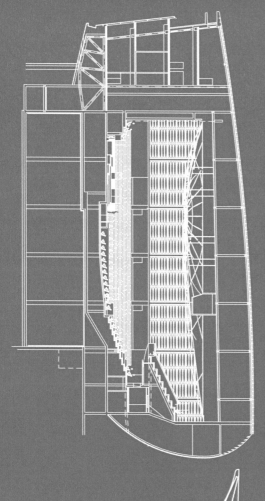

CONCERT HALL – LONGITUDINAL SECTION

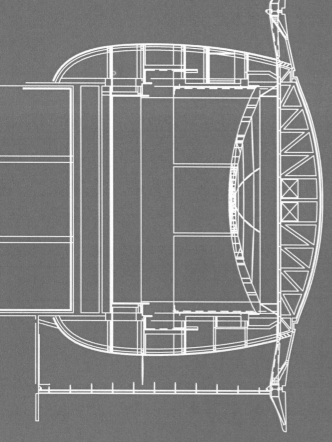

CONCERT HALL – TRANSVERSE SECTION

ACOUSTICAL CONSULTANT:
KIRKEGAARD

ARCHITECT:
NICHOLAS GRIMSHAW PARTNERS
DAVIS BRODY BOND

COMPLETION DATE:
2008

LOCATION:
TROY, NY I USA

CONSTRUCTION TYPE:
NEW CONSTRUCTION

CONSTRUCTION/RENOVATION COST:
$200,000,000

FEATURED SPACE DATA:

ROOM VOLUME:
770,000 FT3

FLOOR PLAN AREA:
13,500 FT2

SEATING CAPACITY:
1,200

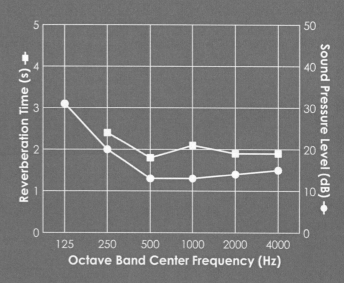

A highly versatile Rehearsal/Recital/Lecture space arose in 2003 within the Claremont Colleges consortium as part of a major and long-awaited expansion and renovation of their music program and a comprehensive renovation and acoustical upgrade of their 700-seat, 25,000sf Garrison Theater. The theater, constructed in 1962, needed to better serve unamplified music concerts. The $12M, 46,000sf project now incorporates the 110-seat, Boone Recital Hall, although the "recital hall" moniker belies its versatility.

Functionally, the MaryLou and George Boone Recital Hall "triples" as a full Choral Rehearsal room (choristers seated as they would be within the Garrison Theater shell), a full-functioning Lecture Hall and, of course, an intimate Recital Hall for soloists and small groups. As a women's college, Scripps encourages choristers especially, so this space features width and sidewall shaping that mimics the orchestra shell in the adjacent Garrison Theater in order that the acoustics of the choral rehearsal environment would "translate" well to the on-stage performance environment in Garrison. To help foster this and provide control over reverberance especially for small recitals, reverberation time adjustment of 0.3s (mid-frequencies) is provided by 642 ft² of heavy drapery at 100% folds, which deploys (from storage) at the outer edge of two substantial cavities that total 6990 ft³ (198 m³).

The little recital hall's significant design challenges included: (a) ensuring the Noise Criterion rating of NC-25 would be met with both the hall's HVAC unit

Interior Unoccupied – Audience | Assassi

immediately behind its upstage wall and the abundant vision glass along a noisy street; (b) ensuring that the vision glass shaping mimicked Garrison's orchestra shell towers; and (c) designing with a ceiling height, and therefore room volume, that was architecturally constrained by exterior geometries.

The full project included, in addition to the Boone and Garrison spaces, nine new music faculty studios, three seminar rooms, fifteen practice rooms within "found" space above the Garrison Stage and below the audience chamber, a music library, and a music classroom. The total project was a great success, acoustically and otherwise.

Opposite page: Choral Rehearsal | Assassi

Interior Occupied – Stage | Conant

Interior Unoccupied – Stage | Assassi

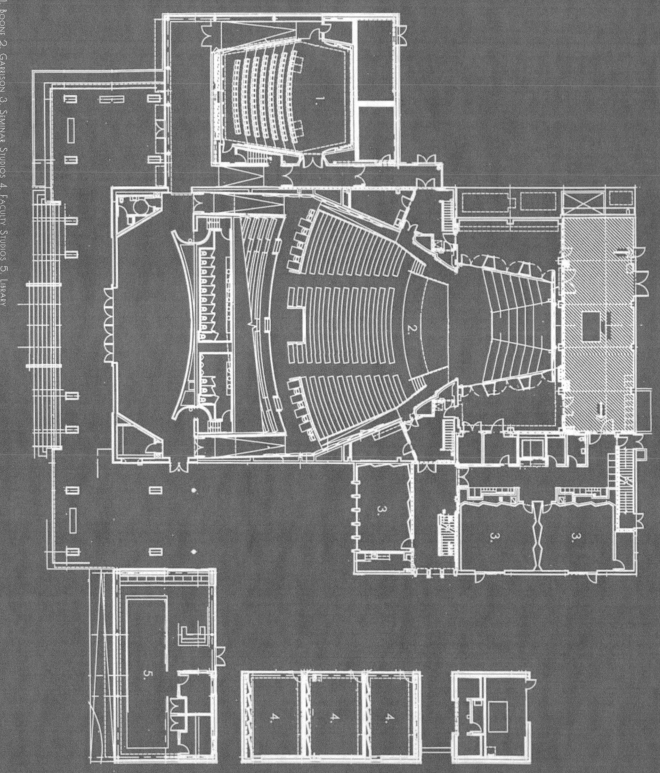

1. BOONE 2. GARRISON 3. SEMINAR STUDIOS 4. FACULTY STUDIOS 5. LIBRARY

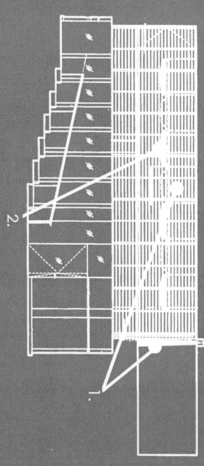

PLAN AND SECTION VIEW

ACOUSTICAL CONSULTANT:
McKAY CONANT HOOVER INC.

ARCHITECT:
BORA ARCHITECTS

COMPLETION DATE:
2003

LOCATION:
CLAREMONT, CA I USA

CONSTRUCTION TYPE:
RENOVATION, EXPANSION

CONSTRUCTION/RENOVATION COST:
$12,000,000

FEATURED SPACE DATA:

ROOM VOLUME:
35,358 FT3

FLOOR PLAN AREA:
1,463 FT2

SEATING CAPACITY:
110

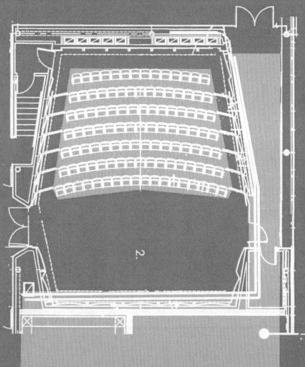

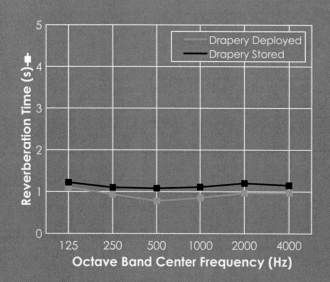

The University of Florida's Music Building Room (MUB) 101 was originally a room dedicated to lectures and band rehearsals. In an effort to increase the functionality of the 4,180 sq. ft. space and serve more students, the School of Music embarked on an extensive renovation plan, seeking a multi-purpose design that would accommodate a variety of functions such as music recitals, lectures and performances which require a richly textured acoustical response in the room. The performances include opera, vocal, string, small natural acoustic groups, and electronic amplified music which requires a very controlled acoustical environment. It was decided to demolish all of the existing interior elements including the floor beside the raked seating area to provide a base for acoustical sculpting that would not only optimize the propagation of musical

sounds to the audience seating area but also for musicians to hear themselves. Removing the existing ceiling also served to increase the acoustical volume of the room.

Under severe budget constraints, designing a room optimized for recitals and performances that required different sound fields for the different musical genres that would be played in the room and which allows sound propagation to occur from multiple sources on the stage to the audience area became an interesting challenge. Accommodating a variety of uses required special space shaping and acoustical elements that can allow for different configurations, size and type of ensemble or genre of performance. Every acoustical element had to be useful and carefully shaped due to cost constraints.

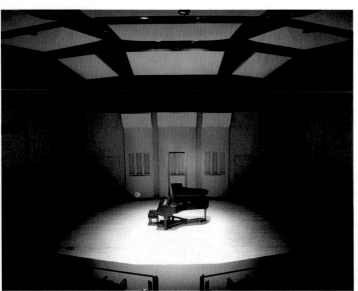

STAGE | SIEBEIN ASSOCIATES, INC.

SIDE WALL | SIEBEIN ASSOCIATES, INC.

VIEW FROM SEATS | SIEBEIN ASSOCIATES, INC.

REAR WALL | SIEBEIN ASSOCIATES, INC.

Acoustical design concepts included: a stage enclosure with movable towers; a carefully shaped ceiling with suspended "clouds" to provide strong, early sound reflections that increase the loudness of sounds heard by the audience; and variable acoustic devices such as retractable acoustical drapes to provide a rich reverberant environment that can be controlled for electronic music, jazz and percussion ensembles as needed. The ceiling above the stage is curved in two directions to allow cross-room diffuse sound reflections so that performers in an ensemble can hear themselves well and their sounds can be reflected to the audience. Shaped side walls include sound diffusing panels to provide lateral sound reflections that enrich sounds by enveloping the audience. Acoustical door and frame assemblies with a vestibule were added to reduce flanking sound

transmission between the room and the rest of the building, which houses other music rehearsal spaces.

The acoustic shell was originally designed to encompass the width of the stage with movable towers that included curved lower areas, QRD-type diffusers in the middle area, and a curved shelf at the top. Due to budget constraints, only the rear portion of the shell was constructed. An interesting feature of the shell is that the middle portion including the QRD area is actually a custom, hidden door that provides access to the rear storage area.

Another component of the remodel project was the addition of a rich, robust, state-of-the-art sound system. Working closely with the building committee and the architect, two stand-alone sound systems were designed that would operate independently of one another: 1) the "house" system to provide support for spoken word during lectures, recording of musical groups, and mild reinforcement for performers who prefer electronic assistance, and 2) design for a very unique separate future system with 16 independent loudspeaker channels for a full immersive sonic experience in experimental electronic music performances. Projection systems with quiet fans that would operate without distracting fan noise and produce crystal clear projection were also implemented.

Acoustical measurements, analysis and design of noise mitigation systems for the HVAC system to provide a quiet environment to appreciate the acoustic subtleties of fine music were also conducted.

Conversations with music faculty and students, and analysis of measurement data revealed that the project was a complete success, providing sound levels and reverberation times that are well suited for both large and small performances of the varied types that will perform in the room. The reverberation time data shown were measured in the unoccupied space. The room is being used for a variety of venues every day of the week and is currently considered the most desirable room to perform in at the university.

In addition to the architect and acoustical consultant, the design team included SK Design and Consulting LLC (theatre consultant).

SHELL | SIEBEIN ASSOCIATES, INC.

STAGE FROM ANGLE | SIEBEIN ASSOCIATES, INC.

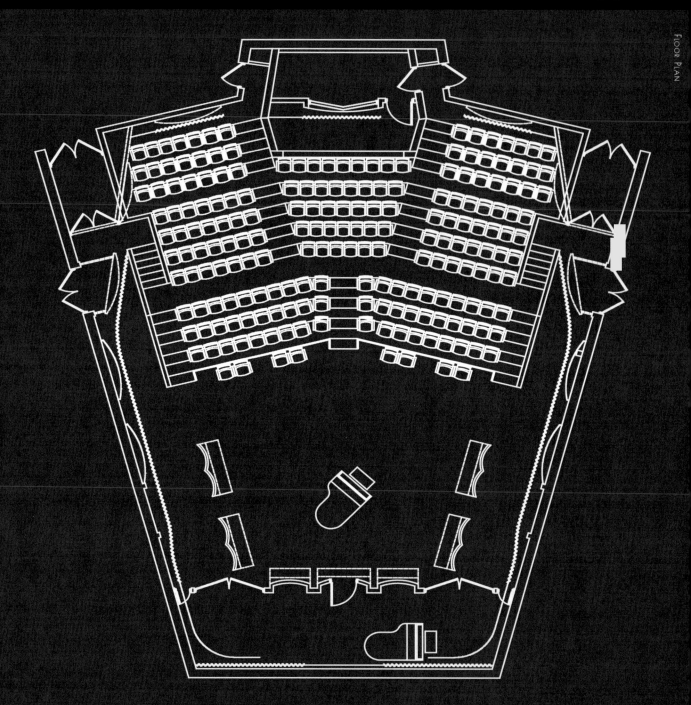

ACOUSTICAL CONSULTANT:
SIEBEIN ASSOCIATES, INC.

ARCHITECT:
BRAME HECK ARCHITECTS

COMPLETION DATE:
2010

LOCATION:
GAINESVILLE, FL I USA

CONSTRUCTION TYPE:
RENOVATION

CONSTRUCTION/RENOVATION COST:
$850,000

FEATURED SPACE DATA:

ROOM VOLUME:
69,605 FT3

FLOOR PLAN AREA:
4,180 FT2

SEATING CAPACITY:
201

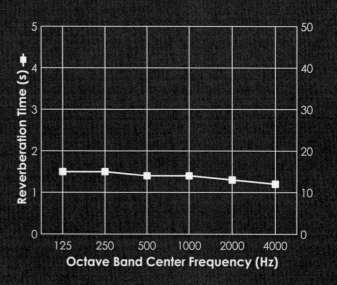

The 17,800 square foot George Steinbrenner Band Building adjacent to the existing music building houses the 5,600 square foot Stephen Stills Rehearsal Hall for the University of Florida "Pride of the Sunshine" Fightin' Gator Marching Band, jazz band, and symphonic band, as well as offices, instrument storage, band library, conference room, and two large lobby areas. The Hall accommodates over 300 musicians and 10 to 12 different ensembles over the course of the academic year as well as through an intense summer season. Renowned musicians and professionals in the music industry have compared the band room's sound field to "world class rehearsal venues and music studios" in the United States and praised the acoustics of the room. Part of this project involved measuring and recording each type of instrument that would typically be present in the different ensembles that will use the room so that optimal space, location, and acoustical response could be provided for each. The room was also used as a case study by Dr. Lucky Tsaih in her doctoral dissertation on the soundscape of music education.

The acoustic design for the band room consisted of a large volume space with a 30 ft high ceiling capable of dissipating the loud sounds when the marching band practices indoors. The room also had to accommodate other large and small ensembles and genres of music ranging from jazz to classical. Retractable acoustic drapes were integrated within the architecture of the room to provide variable acoustics so that specific sound fields can be created to optimize rehearsal and instruction for the many groups that use the room.

The most important acoustical elements in the room are the specially-shaped sound diffusing wall and ceiling panels, which are curved and angled to provide optimal sound reflections to allow the music director to hear each section of the ensemble clearly were carefully coordinated and integrated with the architectural aesthetics of the room. The size, angles, and locations of these elements on the walls and ceiling of the room were initially studied using ray diagrams and later in a 3D computer model of the room so that sound reflection paths between each group of musicians and between the musicians and the instructor could be maintained for the primary configurations of the room. These sound reflection paths are essential for musicians to play in time, in tune, and in dynamics with each other and for the instructor to be able to hear the students while they play. Curved, free standing columns in each corner of the room are shaped to provide low frequency sound absorption and diffusion and also serve as return air ducts. Retractable acoustical drapes that can be visually hidden behind specially shaped columns throughout the room are able to cover the four walls of the room reducing excessive reverberation, to allow variable aural configurations for different ensembles and their seating layouts.

When fully deployed, the drapes provide the optimal acoustical environment for marching band and jazz band rehearsals and brass ensembles by allowing the

South Exterior | Siebein Associates, Inc.

Front Left without Curtains | Siebein Associates, Inc.

band director to identify individuals and small sections that may need more instruction. The drapes can be retracted behind the outer set of angled diffusing surfaces when the wind ensemble, symphonic band, and orchestra use the room, providing a sharp, clean aesthetic. As the stage performance approaches, the drapes are pulled back to more closely simulate the liveliness of a stage environment. Impulse response measurements were made in the room with drapes deployed, for more controlled reverberation while the band plays, and with drapes retracted, for a more lively room. The reverberation time data shown were measured in the unoccupied space.

The acoustical design process included active collaboration with the client, design team, and user groups to develop a tailored system of acoustical elements used to sculpt the sonic qualities of the room from programming through construction administration and post-occupancy. Special details were developed to reduce the sounds from the campus including a large carillon from disturbing rehearsals and recording in the room and specified the sound isolation and mechanical noise and vibration control systems. An integrated approach was taken with the design team to solve complex acoustical challenges including sound isolation and noise and vibration control of the main mechanical room that was located adjacent to the rehearsal room. Custom air handling units, custom elbow silencers, specifically designed sound isolation ceilings in the mechanical room, and a full building expansion joint between the mechanical room and the rehearsal room were successfully designed and constructed to result in a very quiet background noise levels in the rehearsal room. The background noise level data shown were measured in the unoccupied space with the HVAC system on.

In addition to the architect and acoustical consultant, the design team included: Affiliated Engineers SE, Inc. (mechanical engineer) and Parrish-McCall Constructors, Inc. (general contractor).

Above: Rear Wall with Curtains | Siebein Associates, Inc.

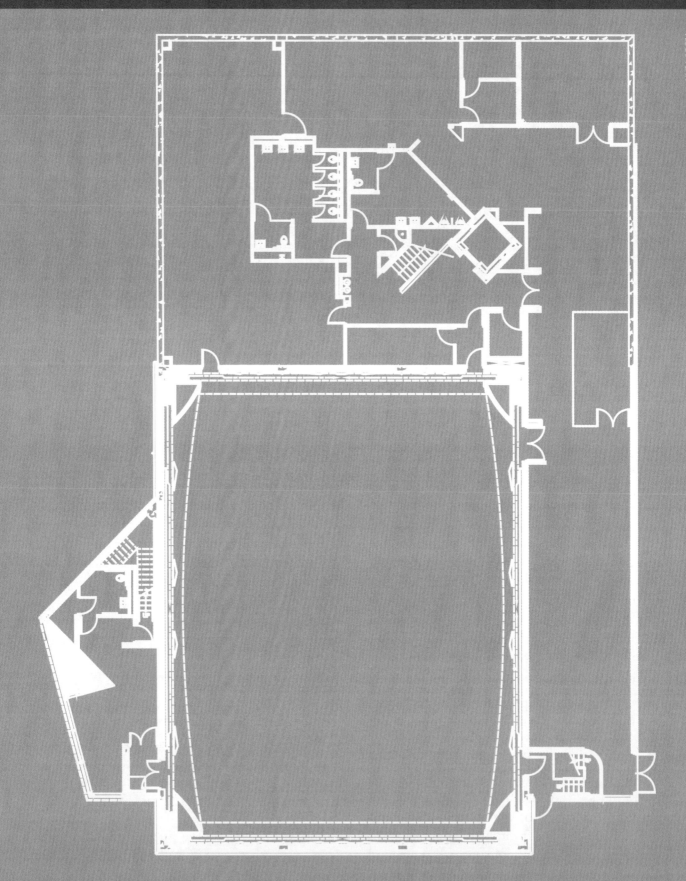

FLOOR PLAN

ACOUSTICAL CONSULTANT:
SIEBEIN ASSOCIATES, INC.

ARCHITECT:
ZEIDLER ARCHITECTURE

COMPLETION DATE:
2008

LOCATION:
GAINESVILLE, FL I USA

CONSTRUCTION TYPE:
NEW CONSTRUCTION

CONSTRUCTION/RENOVATION COST:
$7,884,656

FEATURED SPACE DATA:

ROOM VOLUME:
145,803 FT3

FLOOR PLAN AREA:
5,600 FT2

SEATING CAPACITY:
300

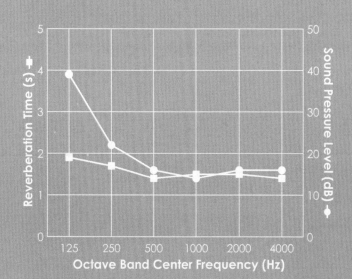

The University of Kansas (KU) School of Music Choral Rehearsal Room 328 is in Murphy Hall on the University campus in Lawrence, Kansas. Murphy Hall was constructed in 1953 and the KU Choral Rehearsal Room is part of the original building. The acoustical improvements described herein were accomplished in 2013. Choral Rehearsal Room 328 is not particularly attractive from an aesthetic viewpoint. As initially constructed, it has several shallow gypsum board triangular wall sections, but the ceiling and floor are both parallel and sound reflecting. Apparently, at some point in time, considerable sound absorbing material was installed in the room, but a previous choral director found this relatively dry room unsuitable for choral teaching and rehearsal and the sound absorbing material was removed. Room 328 was then quite

reverberant with an unoccupied mid-frequency reverberation time (average sound decay time in the 500 and 1000 Hz octave frequency bands) of approximately 3.3 seconds. This situation was called to our attention by a doctoral music student who was enrolled in the School of Architecture course on architectural acoustics and who taught and conducted from time-to-time in the choral rehearsal room. The very reverberant character of the room (**problem 1**) made it difficult to hear individual choral voices by the instructor, for members of a choral group to hear each other, and for the instructor to speak to a choral group.

What is the optimum reverberation time for a choral teaching and rehearsal room? For a room like this it is optimal to have a variable reverberation time that can be changed as needed using adjustable sound absorption. Rooms like the KU Choral Rehearsal Room are ideal candidates for adjustable sound absorption. Since the variable sound absorption can be adjusted separately for each of the four walls, the room's sound reflections can be adjusted to some extent.

The room may be adjusted for the users in the room, which change daily, including different choral groups of varying group size. The different people using the room have different perspectives of what sounds good. The room is also used for various styles of music and audio recording.

Adjustments may also be made to meet the changing needs of periodic events, including vespers, music camp, rehearsal for performance in a performance hall, and placement auditions.

Room adjustments can take place when the faculty or administration changes, since new users may have differing opinions about the room. Adjustability allows flexibility for new people to do what they think is best and can safeguard the room from being moved or removed completely by someone in the future.

Problem 2 was the flutter echo between the parallel sound reflecting ceiling and floor. This flutter was not apparent with the very long mid-frequency reverberation time of 3.3 seconds, but the flutter echo would become apparent and annoying when the

DRAPERY LOWERED 25% LOOKING SOUTHWEST | ROBERT C. COFFEEN

DRAPERY LOWERED AVERAGE 90% | ROBERT C. COFFEEN

reverberation time was lowered by the installation of the adjustable sound absorption.

Problem 1 was solved by the installation of eight vertically rising and lowering sound absorbing drapery banners consisting of two layers of wool surge fabric with approximately 3 inches between layers and with approximately 6 inches between the rear layer and the nearest wall surface. The sound absorbing drapery on each of the four walls is raised and lowered separately so that the sound absorption can be independently established for each wall. The sound absorbing drapery is often extended about 50%.

Problem 2 was solved by installing 4 by 4 feet square and 2 by 2 feet square 3 mm thick thermo molded PVC/acrylic sound scattering pyramidal shapes directly on the gypsum board ceiling. These pyramidal shapes have a mid-frequency coefficient of absorption of approximately 0.03.

The overall shape of the Choral Rehearsal Room could have been improved but funds were not available to change the shape of the gypsum board walls or ceiling, the room lighting, or the location of HVAC return air grilles.

Prior to the installation of the variable sound absorption and the ceiling sound scattering elements the original Murphy Hall HVAC units were replaced with new HVAC machines. This included the air handler serving the Choral Rehearsal Room which is in a mechanical equipment room located behind the Rehearsal Room north wall. The ambient noise source is the HVAC return air grilles. Prior to the HVAC unit replacement, the Rehearsal Room ambient noise was approximately NC-45. Modification to the HVAC return air duct including the installation of a rectangular silencer reduced the Rehearsal Room noise to NC-33. The ceiling supply air linear diffuser does not significantly contribute to the Rehearsal Room HVAC noise. The background noise level data were measured in the unoccupied space with the HVAC system on.

In addition to the architect and acoustical consultant, the design team included: Professional Engineering Consulting (mechanical engineer), KU Design &

Construction Management (general contractor), and AcouStaCorp (furnished and installed AcouStac sound absorbing drapery). Students enrolled in the School of Architecture Acoustic Studio assisted with Choral Rehearsal Room acoustic measurements and assisted with a presentation for the School of Music Dean and the faculty Choral Program Directors.

CEILING SOUND SCATTERING SHAPES | ROBERT C. COFFEEN

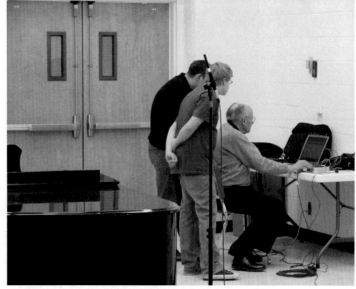

ACOUSTIC STUDIO STUDENTS ASSISTING | ROBERT C. COFFEEN

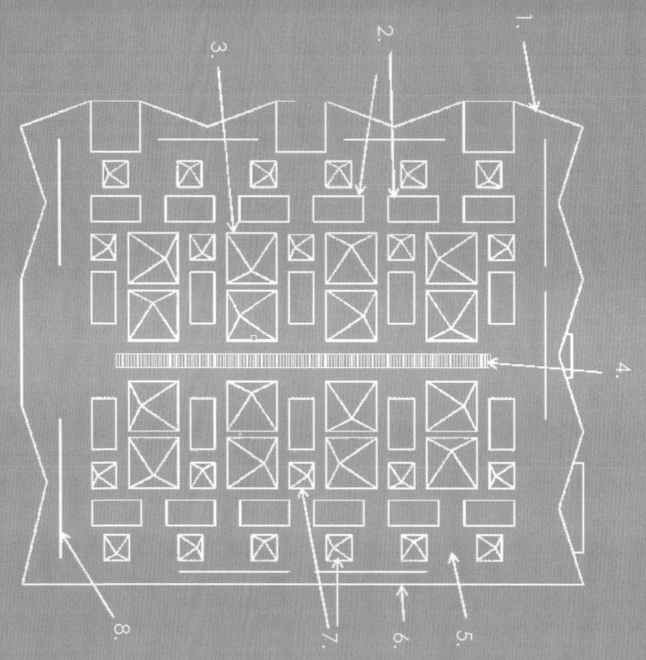

1. Gypsum Wall Board 2. Light Fixtures 3. Large Sound Diffuser 4. Supply Air 5. Vinyl Tile on Concrete 6. Painted Masonry Block 7. Small Sound Diffuser 8. Drapery

ACOUSTICAL CONSULTANT:
THE UNIVERSITY OF KANSAS SCHOOL OF ARCHITECTURE
ACOUSTICS STUDIO

ARCHITECT:
THE UNIVERSITY OF KANSAS DESIGN & CONSTRUCTION
MANAGEMENT

COMPLETION DATE:
2013

LOCATION:
LAWRENCE, KS I USA

CONSTRUCTION TYPE:
RENOVATION

CONSTRUCTION/RENOVATION COST:
$175,000

FEATURED SPACE DATA:

ROOM VOLUME:
27,700 FT3

FLOOR PLAN AREA:
1,628 FT2

SEATING CAPACITY:
125

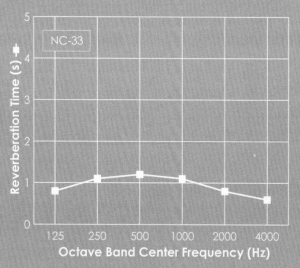

Situated on the north end of the College Park Campus of the University of Maryland, the 318,000 square-foot Clarice Smith Performing Arts Center is a sprawling complex that spreads across 17 acres, forming a "village" that links ten interconnecting buildings serving the University's School of Music and the School of Theatre, Dance, and Performance Studies. The Center is also home to an Artist Partner Program that involves artist residencies, workshops, master classes, and K-12 student events. Serving as a creative catalyst between arts education and the community of College Park, The Clarice was created through the collaboration of the State of Maryland, the University of Maryland, Prince George's County, and many individuals who support art and education. In addition to the architects and acoustical consultant, the design team included Theatre Projects Consultants (theatre consultant).

The Clarice is home to six main performance spaces organized along a grand atrium, a skylit "terraced street." Each venue caters to the specific needs of the program that uses it and at the same time encourages interdisciplinary collaboration. The Dance Theatre incorporates a sprung floor to lessen impact on the dancers' feet, which is beneficial not only to their health but also creates a quieter environment for the instructors and students to hear each other. Retractable seating allows this functional space to be used as both a recital venue and a rehearsal space. Across the hall, the Kay Theatre provides a traditional proscenium stage, complex fly system, and extensive wing space to harbor sets and other production

property.

Music venues are oriented on the other side of the central atrium with the Gildenhorn Recital Hall and Dekelboum Concert Hall directly adjacent to each other. This posed an isolation concern between the two rooms, which was addressed with a number of buffer spaces in the form of a long vestibule serving the concert hall as well as a shared remote recording room.

The 1,096-seat Dekelboum Concert Hall is the largest performance space in the Center. Adjustable elements allow the concert hall to be fine-tuned to the specific needs of contrasting musical ensembles. A system of retractable risers can accommodate instrumental ensembles ranging from a full symphony orchestra to a string quartet. Sound-absorbing banners and draperies offer a degree of acoustical adjustment between unoccupied rehearsals for percussion ensembles and a capella choirs and sold-out performances for jazz/big bands and consorts of early instruments. In lieu of catwalks that would visually obscure the vaulted ceiling, a tension wire grid provides technical access. The hall is designed to support a future pipe organ on the upstage wall above the choral terrace.

The reverberation time data were measured in the unoccupied concert hall with curtains pocketed, and the background noise level data were measured in the unoccupied concert hall with the HVAC system on.

The 156-seat Kogod Theatre is a multi-purpose, flexible black box space often used for performances, meetings, and receptions, while the 86-seat Cafritz Theatre is also a black box theatre used for performances, lectures, meetings and special events. The Performing Arts Library is world-renowned for its extensive archives including the world's largest collection of classical piano repertoire. Modern music is also celebrated with a state-of-the-art Music Technology Lab, which includes 16 workstations featuring synthesizers and an array of software. A more advanced digital workstation is installed in the Computer Music Studio, which is designed for researchers and composers. Twenty-eight additional classrooms, fifty music rehearsal and practice rooms, and one hundred faculty studios and offices round out the facilities. All were carefully designed with respect to room acoustics, sound isolation, and noise and vibration control of building service systems.

Opposite page: Dekelboum Concert Hall | Clarice Smith Performing Arts Center

View of Stage with Choir | Clarice Smith Performing Arts Center Grand Atrium | Clarice Smith Performing Arts Center

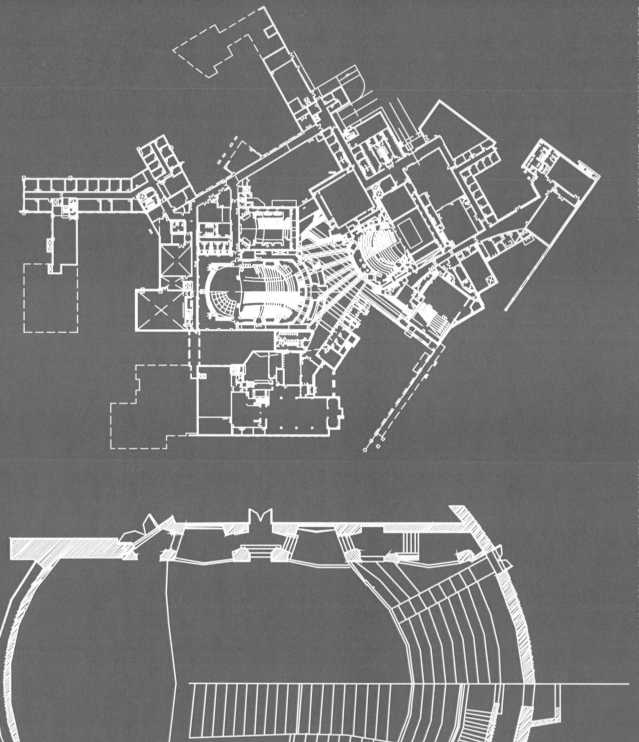

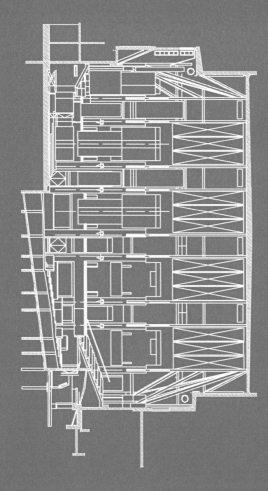

DEKELBOUM CONCERT HALL - LONGITUDINAL SECTION 1

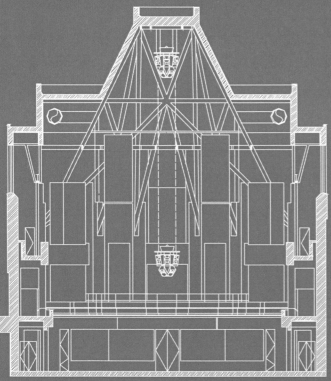

CONCERT HALL - TRANSVERSE SECTION

ACOUSTICAL CONSULTANT:
KIRKEGAARD

ARCHITECT:
MOORE RUBLE YUDELL
AYERS SAINT GROSS

COMPLETION DATE:
2001

LOCATION:
COLLEGE PARK, MD I USA

CONSTRUCTION TYPE:
NEW CONSTRUCTION

CONSTRUCTION/RENOVATION COST:
$86,000,000

FEATURED SPACE DATA:

ROOM VOLUME:
816,000 FT3

FLOOR PLAN AREA:
9,600 FT2

SEATING CAPACITY:
1,096

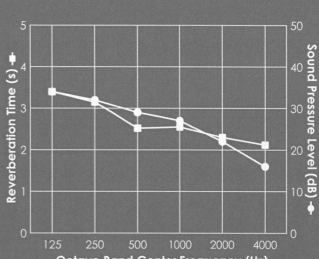

The Hochschule für Musik, or University of Music of Karlsruhe in Germany, opened its doors in 1812 and over the centuries, built its reputation as a center for education and research. As an internationally renowned center in all aspects of music, it is an exciting and stimulating environment in which to work and study. Nowadays, it offers strong curricula in musicological and musical skills to more than 600 students of 50 different nationalities.

Over the centuries, as the number of disciplines and students have grown, so did the range of services offered by the university and the associated requirements for the use of the existing rooms. These requirements could only be fulfilled to a limited extent in the castle (more specifically the Gottesaue Palace), which was the previous prime location for the Hochschule für Musik, as practice rooms were spread all over the city.

Naturally, the need for more suitable spaces for the University ended up in the construction of an extension building called Campus One. This extension to the Hochschule für Musik offers the users a large multi-use hall and all the necessary ancillary and rehearsal rooms (61 teaching and ensemble rooms in total). Campus One is an 'All in One' location, that enhances diversity in music teaching and practice but also gives 'space' to concentration, whether on an individual or collaboration basis. In addition, it accommodates a fully-equipped broadcasting studio for the LernRadio Institute.

As acoustic consultants, Kahle Acoustics worked both for the client to define the acoustic environment

SIDEWALLS AND BALCONIES WITH ACOUSTIC CURTAINS | KFB POLSKA

RETRACTABLE BLEACHERS WITH LOOSE CHAIRS | KFB POLSKA

TILTED WOOD PANELS BELOW THE BALCONIES | KFB POLSKA

(including user equipment and acoustic settings) and for the users. The outcomes were highly flexible spaces. Considering, for example, the large multi-use hall's architecture, one can visualize the fully integrated acoustics. This hall has a 450- to 550-seat capacity and can function as an opera house. It includes an orchestra pit and raked seating that can accommodate an audience up to 550 seats when used for concerts and recitals. The stage house is black with lighting bridges, and there are manual curtains for variable acoustics. These manual curtains can very easily be hung and moved around not only by the technical director but by the students too. When specific installations are required, the telescopic seating platforms are retracted. One can then use the hall's sidewalls in order to modify acoustics as their wood-slatted acoustic surfaces below the balconies can be slightly tilted.

The large multi-use hall offers a generous and balanced reverberation that provides richness to the orchestra sound with a very good response in bass frequencies. The reverberation time data were measured in the unoccupied space with the curtains down. The sidewalls provide most of the early energy and therefore create a feeling of spaciousness. The acoustic homogeneity as a whole is based on a well-optimized geometry making the sound very clear and rich despite the width of the hall. The noise ratings were calculated for the unoccupied space.

Campus One's multimedia complex is among the largest and liveliest best-equipped music institutions in Germany. It stands as an exciting and stimulating environment in which to work and study music with a wide variety of perspectives.

Opposite page: Stage Area | KFB Polska

1. Concert Hall 2. Multimedia Room 3. Classroom

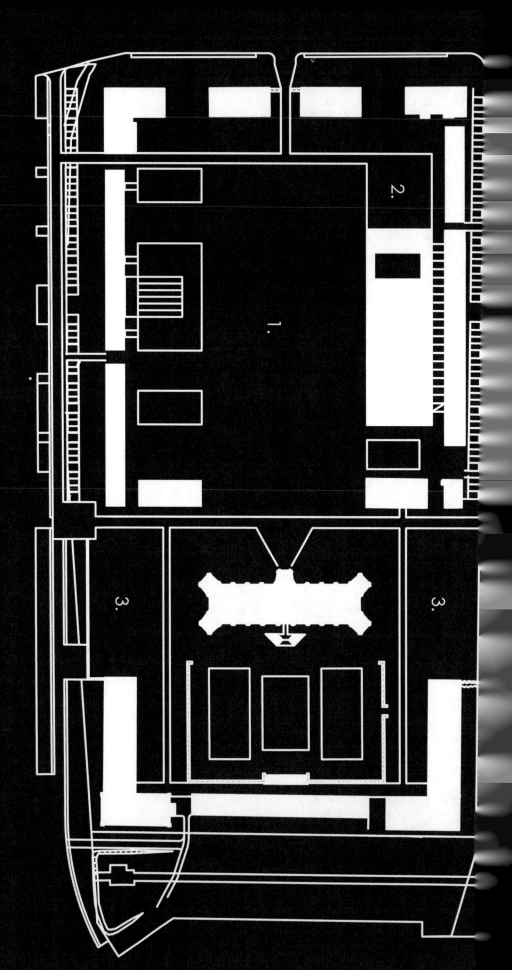

ACOUSTICAL CONSULTANT:
KAHLE ACOUSTICS

ARCHITECT:
ARCHITEKTEN.3P FEUERSTEIN RÜDENAUER & PARTNER /
ARCHITEKTURBÜRO RUSER & PARTNER

COMPLETION DATE:
2013

LOCATION:
KARLSRUHE I GERMANY

CONSTRUCTION TYPE:
NEW CONSTRUCTION, EXPANSION

CONSTRUCTION/RENOVATION COST:
$32,840,250

FEATURED SPACE DATA:

ROOM VOLUME:
233,077 FT3

FLOOR PLAN AREA:
7,104 FT2

SEATING CAPACITY:
450 (OPERAS), 550 (RECITALS AND CONCERTS)

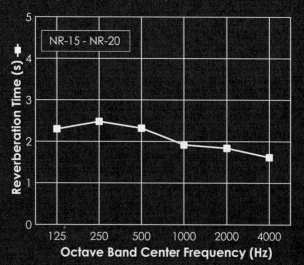

This new 122,000 square-foot home for the School of Music at the University of Tennessee caters to 21st Century needs of students, faculty, and guest artists. A 400-seat recital hall serves as the facility's centerpiece and accommodates student and faculty recitals and chamber music performances. The building also includes a dedicated choral rehearsal room, opera orchestra rehearsal room, practice rooms, music teaching studios, a recording mixing lab and the 7,000 square-foot WJ Julian Marching Band Rehearsal Room that accommodates 350 musicians.

To control loudness in the marching band rehearsal room, the underside of the roof deck is located 48 feet above the floor, with suspended ceiling elements located 30 feet above the floor. The Marching Band rehearsal room includes adjustable acoustics in the form of hinged panels on the lower side walls, motorized acoustical curtains on the upper walls below the suspended ceiling elements, and a third set of motorized curtains that can be extended over the walls between the underside of the roof and the suspended ceiling elements. There are also fixed sound absorbing wall panels to supplement these adjustable elements. The wide range of liveness allows this room to also serve as a rehearsal room for many other ensembles, increasing this space's utilization. During marching band rehearsals, it is common to have the curtains extended on the middle level and the hinged panels opened to maximize sound absorption.

The new facility was constructed on the same site as the previous smaller music building. Because of the

RECITAL HALL – REAR VIEW | ACOUSTIC DISTINCTIONS

MARCHING BAND REHEARSAL ROOM | ACOUSTIC DISTINCTIONS

BUILDING EXTERIOR | ACOUSTIC DISTINCTIONS

limited available footprint for the new building, many of the acoustically critical spaces had to be vertically stacked, which added to the challenge of achieving adequate sound isolation between spaces. Part of the solution was to construct the facility as three separate buildings.

The acoustics of the recital hall are adjustable to support a wide range of music genres including jazz, acoustic recitals, choral and instrumental chamber music. There are motorized, adjustable acoustic curtains that ring the room on three levels. On the two levels in view to the audience, there is an acoustically transparent wood grille so that the visual appearance of the room is consistent, regardless of whether the curtains are extended or retracted. The reverberation time data were measured in this unoccupied space along with the background noise level data, which were measured with the HVAC system on. It is worth noting that the reverberation does vary with the adjustable design.

The mechanical system is a virtually silent, simple overhead air distribution system consisting of open-ended ducts. Return air is pulled out from underneath the stage platform.

In addition to the architect and acoustical consultant, the design team included: Acoustic Distinctions, Inc. (audio systems designer, video systems designer), I.C. Thomasson Associates, Inc. (mechanical engineer), and Johnson & Galyon Construction (general contractor).

Opposite page: Recital Hall - Front View | Denise Retallack

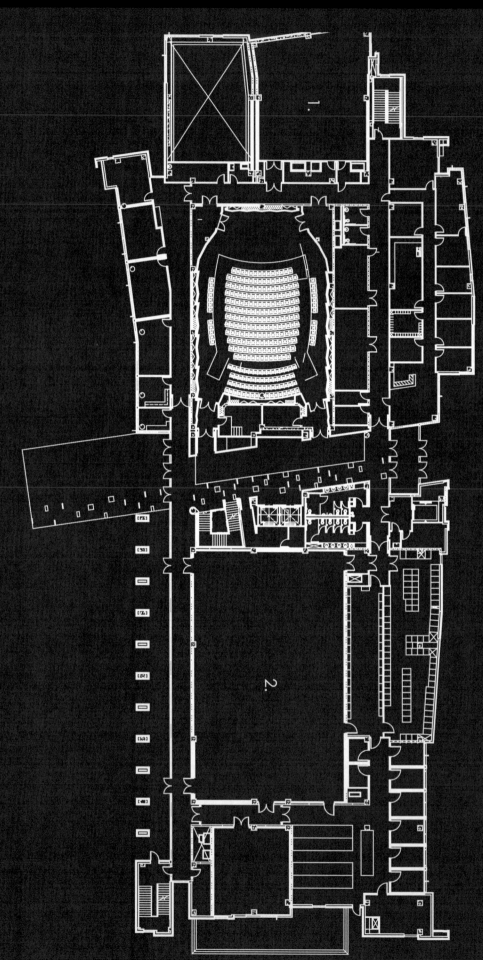

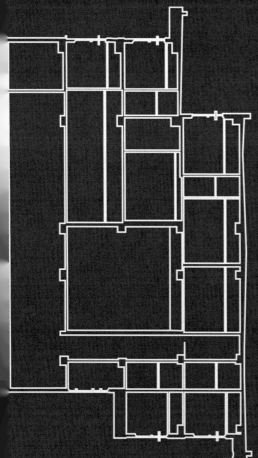

LONGITUDAL SECTION - EAST-WEST

SECTION - NORTH-SOUTH

ACOUSTICAL CONSULTANT:
ACOUSTIC DISTINCTIONS, INC.

ARCHITECT:
ASSOCIATED MUSIC CENTER ARCHITECTS
(JOINT VENTURE OF BARBARA MCMURRY
ARCHITECTS AND BLANKENSHIP & PART-
NERS)

COMPLETION DATE:
2013

LOCATION:
KNOXVILLE, TN | USA

CONSTRUCTION TYPE:
NEW CONSTRUCTION

CONSTRUCTION/RENOVATION COST:
$35,000,000

FEATURED SPACE DATA:

ROOM VOLUME:
484,000 FT3

FLOOR PLAN AREA:
6,520 FT2

SEATING CAPACITY:
402

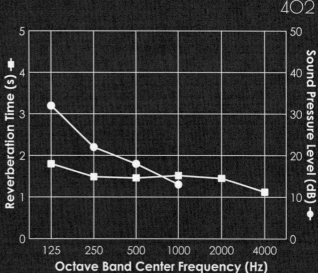

ocated in a popular park-like area of the Wenatchee Valley College campus in Wenatchee, Washington, the new Music and Arts Center (MAC) was thoughtfully designed to reflect the programs housed within. Shared by the music and art programs, the MAC provides much-needed art studio space and state-of-the-art, acoustically-designed music rehearsal and performance spaces.

The building's interior requirements are diverse and specific to enhancing the education missions of each department. The MAC is organized into two distinct wings – Music to the south and Art to the north – with a shared student gallery and lobby area that provides an important, central and transitional "pivot point" connecting both wings while becoming the heart of the building.

The music wing includes a 150-seat recital hall, rehearsal space, recording studios, classrooms, and practice and support areas. The art wing includes painting, ceramics, sculpture and 2-D design/print-making studios, graphic design studios, classrooms, and exhibit and support areas. With a sculptural glass and steel entry, the "pivot point" connects the wings and provides an element of hierarchy for the building. An expansive glass wall adjacent to a covered art patio provides views to the historic Well's House to the north.

Needing natural light in the art wing, north light is brought into the studios by utilizing large double-height windows. The appropriate placement of windows provides much-needed natural daylighting for the studios while minimizing summer cooling

RECITAL HALL GILLS | LARA SWIMMER PHOTOGRAPHY

NORTH ELEVATION / ART STUDIOS | LARA SWIMMER PHOTOGRAPHY

LOBBY | LARA SWIMMER PHOTOGRAPHY

loads. Versatility is provided in two art studios through the use of overhead garage doors that allow seasonal expansion and access to the outdoors while capturing views of the campus to the north. The music wing recital hall features acoustically shaped "gills" that reflect sound waves and colored light at the west wall. The reverberation time data were measured in this unoccupied space. The building strives to be sculptural and an object of art itself with simple brick planes, metal panels, and hot-rolled steel siding at the student gallery that will acquire a patina over time, adding more tonal value and texture to this compositional element.

The greatest design challenge was separating the art and music wings (because of unique program needs and diverse system requirements) while preserving the prized setting. The building was strategically sited between valued specimen trees and great care was taken to remove as few trees as possible. For every tree that was removed, three were planted. Overall, the landscape is intended to blend and complement this park-like area of campus while utilizing native grasses and drought tolerant plants in specific areas to minimize irrigation.

Sustainable design initiatives were intentionally incorporated into the design even though LEED certification was not a high-priority for the owner. As an inherent part of the designer's philosophy, the issues of sustainable design were at the forefront and resulted in an energy-efficient building designed to LEED Silver standards.

In addition to the architect and acoustical consultant, the design team included: PLA Designs (theatre consultant), Kirkegaard (audio systems designer, video systems designer), Escent Lighting (lighting designer), MW Consulting Engineers (mechanical engineer), Graham Construction (general contractor), Integrus Architecture (interior design, structural engineer), Roen Associates (cost estimating), GeoEngineers, Inc. (geotech engineer), Coffman Engineers (civil engineer), and SPVV Landscape Architects (landscape architect).

Opposite page: Recital Hall | Lara Swimmer Photography

1. RECITAL HALL. 2. REHEARSAL. 3. RECORDING STUDIO. 4. PRACTICE ROOM.

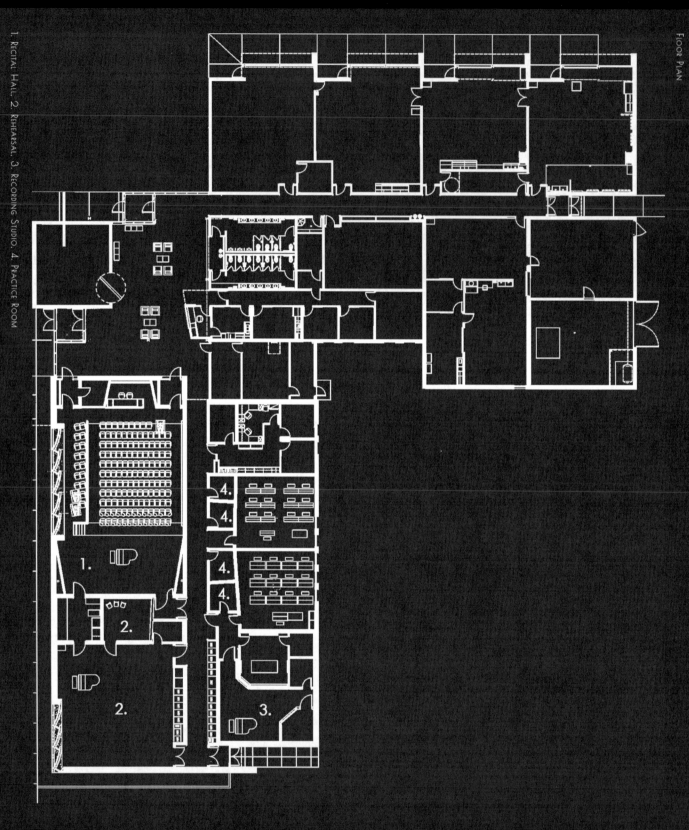

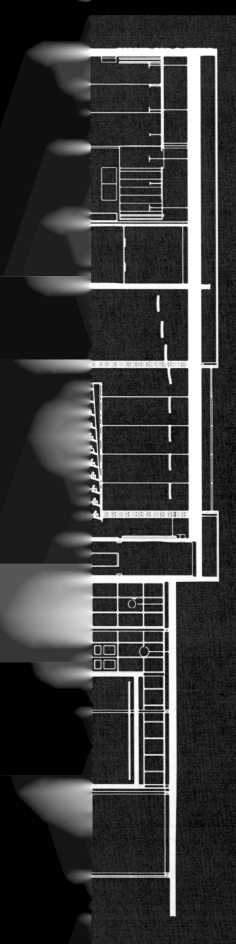

SECTION

ACOUSTICAL CONSULTANT:
SPARLING, A STANTEC COMPANY

ARCHITECT:
INTEGRUS ARCHITECTURE

COMPLETION DATE:
2012

LOCATION:
WENATCHEE, WA I USA

CONSTRUCTION TYPE:
NEW CONSTRUCTION

CONSTRUCTION/RENOVATION COST:
$6,800,000

FEATURED SPACE DATA:

ROOM VOLUME:
75,868 FT3

FLOOR PLAN AREA:
1,885 FT2

SEATING CAPACITY:
158

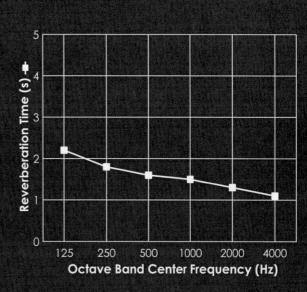

Western Connecticut State University's new School of Visual and Performing Arts facility supports students, faculty, and the community. Jaffe Holden collaborated with the full design team in addition to department chairs in art, music, and theatre, so the new facility would meet the needs of all end-users. Specifically, the design team included: Fisher Dachs Associates (theatre consultant), Jaffe Holden Acoustics (audio systems designer, video systems designer), Fisher Dachs Associates (lighting designer), Kohler Ronan Consulting Engineers (mechanical engineer), and Dimeo Construction Company (general contractor). The center includes a proscenium theatre, concert hall, recording studio, live room, music classrooms, jazz and choral rehearsal rooms, dance studios, art studios, an art gallery, black box theatre, faculty studios, and practice rooms.

A new 350-seat concert hall has clear and resonant acoustics that accommodate the full sonic power of a symphony orchestra despite a small seating occupancy. The design team did not want reflectors in the hall to be visual distractions, so the acoustical consultants analyzed the geometry of the room to use the ceiling and bare-pit walls to reflect sound to musicians in lieu of reflectors. Curved stage walls were elongated and repeated to provide diffusion to blend sound and early reflections to enhance onstage hearing.

The height of the room is quite tall. The ceiling material is thick wood plank with applied half round timber logs to provide further diffusion. A series of motor-operated, computer-controlled, and adjustable

Concert Hall – Stage | Jaffe Holden Acoustics

Concert Hall Banners – Deployed | Jaffe Holden Acoustics

Concert Hall – Audience | Jaffe Holden Acoustics

Large Choir on Stage | Jaffe Holden Acoustics

PROSCENIUM THEATER | JAFFE HOLDEN ACOUSTICS

RECORDING STUDIO | JAFFE HOLDEN ACOUSTICS

WOOD DIFFUSING PANELING | JAFFE HOLDEN ACOUSTICS

bottom-up banners can be deployed during amplified performances to absorb sound and cut off the upper side volume of the hall in order to reduce the room's resonance. These banners are housed behind the seating level of the upper ring and deploy vertically up to the roof line.

The reverberation time data were measured in the unoccupied concert hall. The background noise level data were measured in the unoccupied concert hall with the HVAC system on, and the corresponding noise ratings were calculated.

All studios and practice rooms have been designed with acoustic isolation construction methods to ensure the proper amount of sound isolation exists between rooms. The mechanical systems have been designed to achieve the appropriate background sound level. All of the rehearsal rooms have adjustable wall panels, so the room can be tuned to the appropriate acoustics to support the specific group. The recording studio's live room incorporates a very diffusive ceiling made of edge grain wood stalactites, and the walls have sliding wood panels that reveal acoustical panels to adjust the room acoustics. Finally, there are two sound isolation booths off of the live room to allow for proper drum or voice close miking.

The theatre features an industry-standard theatrical sound system that prepares students for careers in live theatre. Computer-based multi-track playback, digital mixing, and multi-channel speaker systems are core components of the system. The audio/video system in the concert hall uses video projection and sound reinforcement for performances on stage. Motorized microphones in the concert hall can be controlled remotely from the recording control room. The recording studio suite is equipped with the latest in analog and digital technologies and can record live performances in both spaces, as well as mix them down into stereo or surround formats. Direct tie-lines between the live room, the control room, and all rehearsal and performance rooms allow live or archival recording.

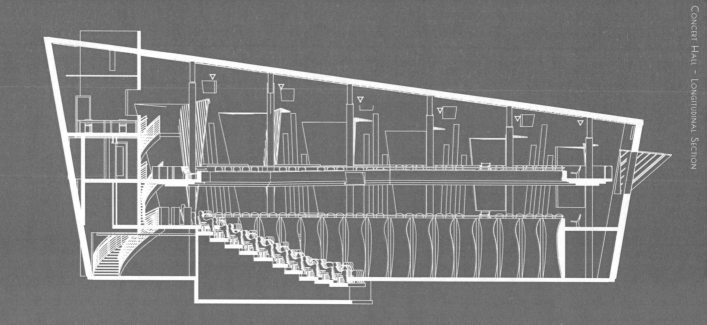

Concert Hall – Longitudinal Section

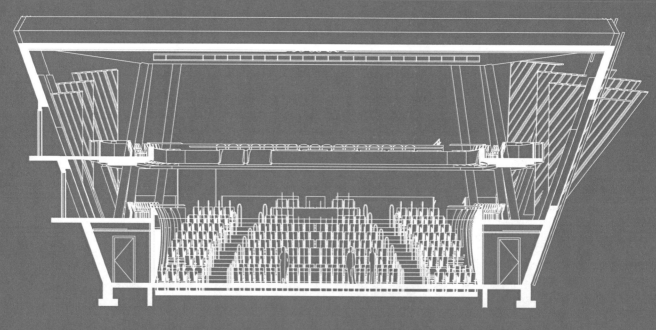

Concert Hall – Cross Section

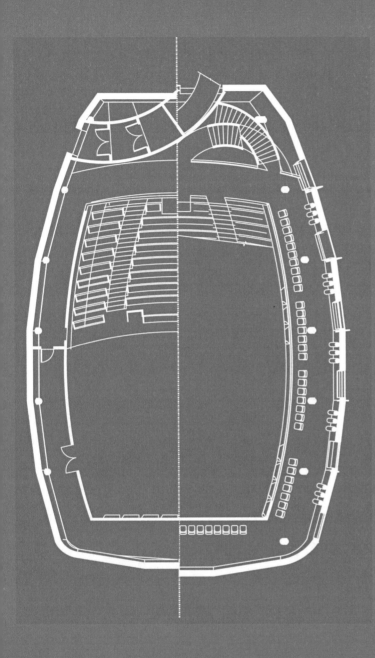

CONCERT HALL PLAN

ACOUSTICAL CONSULTANT:
JAFFE HOLDEN ACOUSTICS

ARCHITECT:
HOLZMAN MOSS BOTTINO / AMENTA
EMMA ARCHITECTS

COMPLETION DATE:
2014

LOCATION:
DANBURY, CT I USA

CONSTRUCTION TYPE:
NEW CONSTRUCTION

CONSTRUCTION / RENOVATION COST:
$97,500,000

FEATURED SPACE DATA:

ROOM VOLUME:
275,000 FT3

FLOOR PLAN AREA:
8,400 FT2

SEATING CAPACITY:
350

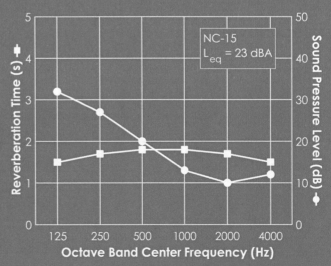

The Marion Buckelew Cullen Center was recently constructed in 2014 for vocal instruction, rehearsal, and performance at the Westminster Choir College in Princeton, New Jersey. The centerpiece of the Cullen Center is the grand Hillman Performance Hall.

A highly successful dual-purpose space, Hillman Hall is used for rehearsals of large choral groups, accommodating up to 235 voices, and for solo and small group recitals. This 3,415 square-foot space features seating on fixed risers, which are used by singers for large choral rehearsals and by audience for recitals. The risers are arranged roughly in a semi-circle around a central area occupied by the choir director or performer, piano and accompanist, and teaching wall. The space extends more than 30 feet

up to the underside of a peaked roof structure. A system of suspended reflectors provide supporting early reflections among choral sections and between the director and chorus or performer and audience. A system of operable curtains located behind the risers allows users to reduce the reverberation time (RT_{mid}, unoccupied) from 1.7 seconds to 1.2 seconds. Deploying only the rear or side curtains leads to reverberation times of 1.4 seconds and 1.5 seconds, respectively.

The dual-purpose design is central to the pedagogy. In a teaching context, students switch between the roles of choral singers and audience sometimes multiple times during a single class. During a choral rehearsal or class, instructors often demonstrate techniques or invite individual students or sections

to present vocal excerpts to the larger group. The operable curtains allow users to reduce reverberation for clarity and close listening during rehearsal or to increase reverberation for a lush room response appropriate to performances.

The building layout protects the hall from other potentially noisy spaces, avoiding the need for specialty sound isolating constructions. Access to the hall is by way of two vestibule entrances located within a series of storage and other rooms that buffer the hall from the lobby.

A critical component of the hall's success is the very low background noise achieved, which was measured to be several points below the NC-15 rating and was below the limits of typical measurement equipment in the upper frequency bands. The unweighted background noise level data were measured in the unoccupied space with the HVAC system on and include instrumentation noise in the upper frequency bands. A dedicated air handler located in a nearby mechanical room delivers supply air to the upper volume of the hall at a low velocity (240 feet-per-minute at open duct ends) via a self-balancing system of symmetrically branching ductwork. No volume dampers are needed within the hall. Return air exits

the hall through two large louvered wall openings to full-height plenum spaces in the front corners of the hall.

In addition to the architect and acoustical consultant, the design team included: Acentech Incorporated (audio systems designer, video systems designer), The Lighting Practice (lighting designer), Schiller and Hersh Associates (mechanical engineer), and Harrison-Hamnett, PC (structural engineer).

The successful design was made possible by the high level of engagement and expectation on the part of the school, who repeatedly pressed the design team to maintain focus on the high quality and function of Hillman Hall and to achieve it within a relatively modest budget of $8.5 million.

Opposite page: Hillman Performance Hall | Halkin/ Mason Photography

Above: Music Classroom | Halkin/Mason Photography

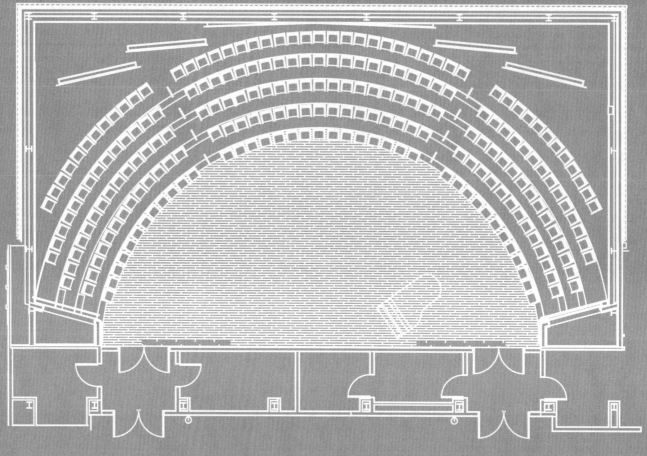

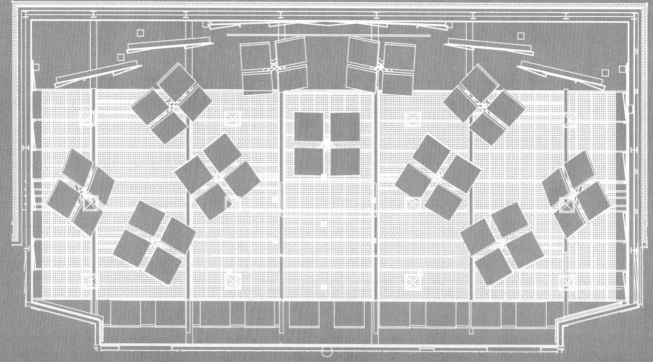

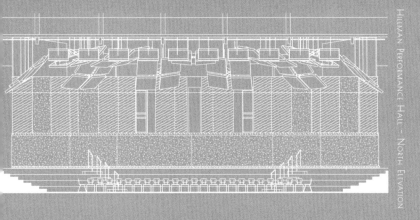

Hillman Performance Hall – North Elevation

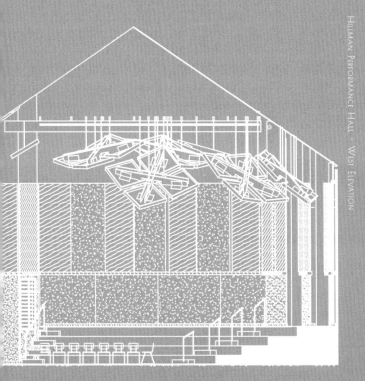

Hillman Performance Hall – West Elevation

ACOUSTICAL CONSULTANT:
ACENTECH INCORPORATED

ARCHITECT:
KSS ARCHITECTS

COMPLETION DATE:
2014

LOCATION:
PRINCETON, NJ | USA

CONSTRUCTION TYPE:
NEW CONSTRUCTION

CONSTRUCTION/RENOVATION COST:
$8,500,000

FEATURED SPACE DATA:

ROOM VOLUME:
106,000 FT3

FLOOR PLAN AREA:
3,415 FT2

SEATING CAPACITY:
250

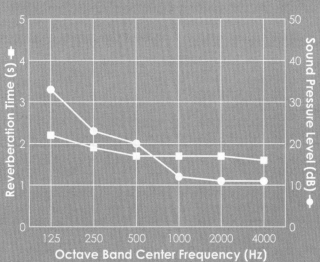

n the last century, Cincinnati's two Catholic, gender-segregated institutions of higher learning were Xavier University in Avondale and Our Lady of Cincinnati in Walnut Hills. Xavier University's first four academic buildings (Hinkle, Alumni, Schmidt and Abbers) were designed by prominent Cincinnati architects, Joseph and Bernard Steinkamp, and built between 1920 and 1929. They are patterned along the lines of the tenth-century Castle of Xavier, located in Javier, Navarre, Spain, which was the birthplace and childhood home of St. Francis Xavier.

Changing society thinking of the late 1960's led both colleges to turn co-ed, and Our Lady of Cincinnati changed its name to Edgecliff College in 1970 to reflect that change. Challenging economic pressures of the late 1970's encouraged the much larger Xavier University to acquire Edgecliff College, helping to save the smaller college's programs in art, music, nursing, and social work. Under the leadership of the late Dr. Helmut Roehrig, Edgecliff's Department of Music became part of Xavier's College of Arts and Sciences. Under Maestro Roehrig's vision, the department developed a B.A. degree in Music Performance and a B.S. degree in Music Education. With the financial crisis averted, all academic departments on the Edgecliff campus were moved into Alumni Hall on the Xavier campus, and the Edgecliff campus was sold in 1987 to prepare for future growth.

Plans were developed in the late 1990's to renovate Alumni Hall into a music education facility and rename it as Edgecliff Hall. This project would replace lost Edgecliff campus facilities and help students and alumni to memorialize Edgecliff College life. Engineers were tasked with planning the complete tear out and renovation of the building and its systems while maintaining Steinkamp's historic facades. The late concert pianist and music department chair, Dr. Dona Buel, led a team of faculty members in designing the function and acoustics of each space.

Faculty team members were interviewed to determine their acoustics preferences. These preferences were then used to create FIR filter-generated acoustic simulations of each space to help the team find consensus, if not agreement, among their differing perceptions. For example, the team members had interestingly divergent points of view for the desired acoustics of Long Recital Hall (named in memory of Edgecliff graduate and choir instructor, Paulina Howes Long). Dr. Buel and others on the team favored a hall with minimal acoustic support to challenge and develop strong performers. Other team members favored a hall with fully supportive acoustics. Dr. Roehrig argued for acoustics that could be adjusted between the two extremes. The compromise was a recital hall having a moderate, flat reverberation spectrum with few early reflections. Some adjustment is possible by the positioning of curtains covering the large, glass walls on each end of the hall.

Long Recital Hall's dark walnut, raised panel walls were milled locally by Hyde Park Lumber and its maple hardwood stage floor was milled locally by

EXTERIOR FACADE RECALLS XAVIER CASTLE

FINISHING STUDIO FLOATING FLOOR

Robbins Sports Surfaces. Both serve to mellow the acoustics of the hall while recalling the ballroom finishes of Maxwelton Hall, the original site for recitals on the old Edgecliff campus. Large bowl, pendant light fixtures suspended from deep ceiling coffers complete the ballroom look. The ceiling was assembled using Echophon Focus E tiles in a Connect grid with coffers formed using S-line internal grid corners and hand-trimmed tiles. The large ceiling surface provides the hall's primary sound absorption and easy access to the mechanical, electrical, and plumbing systems hidden above.

The Echophon system used in Long Recital Hall was also used in the Instrumental Rehearsal Studio to form a three-tiered ceiling. Here, the orchestra arrangement was flanked by two large, glass, curtained walls, while both the front and rear walls were reserved for lockers and whiteboards. Tiering the ceiling was a way to introduce vertical absorbing surfaces facing each direction of the room, despite the lack of available wall space which would typically be used for that purpose.

The Robbins hardwood sports floor used in Long Recital Hall was also used in the Dance Studio located on the second floor. To help prevent footfall noise from reaching the recital hall below, the maple hardwood floor was floated on a two-inch Kinetics Noise Control RIM roll-out floor isolation system. Sound absorption on the walls is provided by 48 millimeter Sonex Valueline panels framed in wood trim and placed out of reach within an upside-down wainscoting. The Sonex treatment used in the dance studio also appears on the side and front walls of the Piano Lab; a tiered lecture room designed for keyboard instruction.

By far, the most striking acoustic surfaces within Edgecliff Hall are the custom-made, binary diffusion walls in the recording suite and within each of the individual practice rooms located in the basement. The idea for the surface is loosely based on the maximum-length binary sequences developed by Schröder [1] for diffuse sound reflection. The local mill produced the panels as prefinished, pseudo-random patterns of one-, two- and five-inch wide by one-inch thick dimensioned lumber attached to

oriented strand board stock. Once the panels were attached to the walls, finishing involved filling the spaces with field-cut, sound-absorbing material. Originally envisioned with gold fabric-covered fiberglass to complement the walnut wood stain, last-minute cost and labor savings favored white melamine foam as a substitute.

In addition to the architect and acoustical consultant, the design team included Messer Construction (general contractor). The reverberation time data and noise ratings were calculated for the unoccupied space.

[1] Schröder, Manfred R. "Diffuse Sound Reflection by Maximum-Length Sequences." *The Journal of the Acoustical Society of America*, vol. 57, no. 1, 1975, pp. 149-150.

RECORDING STUDIO CONSTRUCTION

PIANO INSTRUCTION LAB | JOSHUA L. MOONEY

XAVIER UNIVERSITY

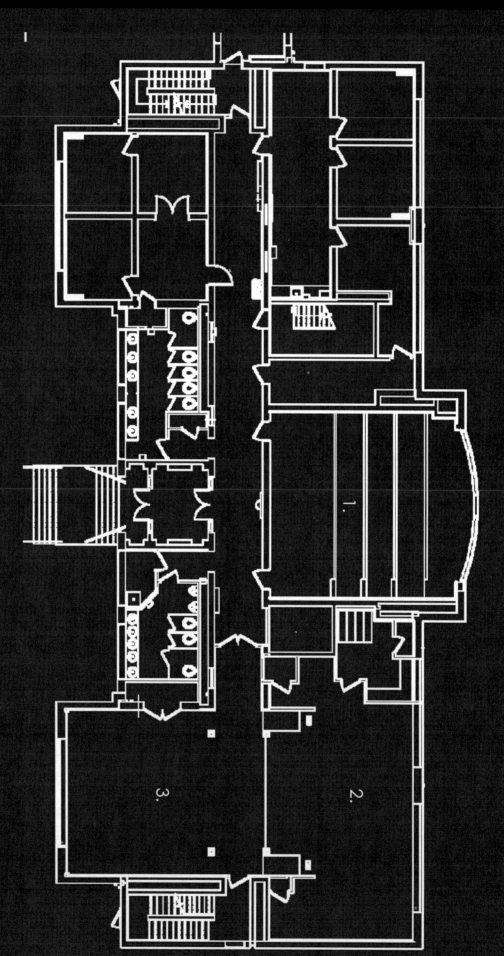

1. LECTURE HALL 2. STAGE 3. LONG RECITAL HALL

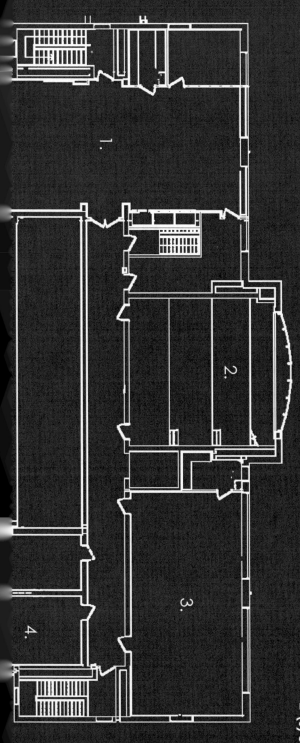

SECOND FLOOR PLAN

ACOUSTICAL CONSULTANT:
ACOUSTICS BY JW MOONEY

ARCHITECT:
MCGILL, SMITH, PUNSHON, INC

COMPLETION DATE:
2000

LOCATION:
CINCINNATI, OH | USA

CONSTRUCTION TYPE:
NEW CONSTRUCTION, ADAPTIVE REUSE

CONSTRUCTION/RENOVATION COST:
$3,000,000

FEATURED SPACE DATA:

ROOM VOLUME:
16,000 FT3

FLOOR PLAN AREA:
1,800 FT2

SEATING CAPACITY:
100

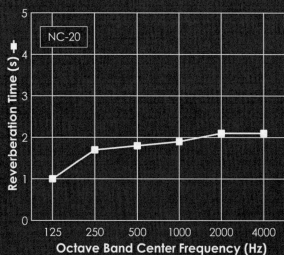

L. M. Ronsse et al. (eds.), *Rooms for the Learned Musician*

HIGHER EDUCATION FACILITIES:
COMPLETED 2015-2020

Carleton College had purchased a high school building, located on the edge of their campus, which was being used by the theatre, dance, and cinema and media studies departments. At the same time, the existing music building, in addition to having acoustical challenges, was of poor quality and was in need of replacement. Furthermore, some of the music teaching studios were located in a separate building that was not designed for music, which resulted in poor sound isolation between music teaching studios and classrooms, both horizontally and vertically. The college's vision to put all of the performing arts and media studies into one building resulted in this project, a $28M dollar expansion and establishment of the Weitz Center for Creativity.

The design of the concert hall was unusually challenging because only 400 seats were needed, yet the room needed to sound like a large concert hall for large wind ensembles and a full orchestra. Another design challenge was to support the college's vision for providing stronger collaborations between the various performing arts disciplines. The hall was designed to support dance and generally higher production values than would normally be provided in a concert hall.

Acoustic Distinctions, Inc. (AD) was engaged to provide acoustics and AV system consulting and design. The effort entailed expanding the southeast corner of the Weitz Center to provide performance spaces, rehearsal spaces, teaching studios and practice rooms. Rather than provide traditional spaces such as an instrumental rehearsal room, a choral rehearsal room, etc., large, medium, and small rehearsal rooms were created. The "large" rehearsal room is essentially the rehearsal room for large instrumental ensembles and has some but limited adjustability. The "medium" rehearsal room is essentially a recital hall and choral rehearsal room. The "small" rehearsal room is mostly for chamber music rehearsal.

The site was very constrained, which resulted in some challenging horizontal and vertical adjacencies. Despite these challenges, ample height and room volume were provided in the large and medium rehearsal rooms, the latter being primarily a choral rehearsal room that doubles as a recital hall.

The concert hall's room volume was maximized despite the site constraints and limited plan dimensions to enhance visual intimacy for dance and other uses. However, a wide range of reverberation was available in the concert hall to accommodate both a theatre/dance mode and a concert hall mode. The reverberation time data were measured in the unoccupied space.

AD's team provided design criteria and maintained involvement through an integrated design approach in addressing needs to achieve strong performance and learning environment for:

- 400-seat performance / concert hall – high acoustic quality; able to support staged dance productions
- Center's interdisciplinary functionality for collaboration among art disciplines and other curriculum areas
- Large instrumental rehearsal room
- Mid-size rehearsal room / recital space; flexible to accommodate varying sized groups and function as small performance and master class spaces with seating for less than 100
- Small rehearsal space for chamber music
- Offices / music faculty studios and small practice rooms

A key feature of the acoustical and architectural design of the new Kracum Performance Hall is patterned wood grilles that accomplish two goals. First, these grilles allow sound to pass through to reach retractable, sound-absorbing curtains. Second, they create a memorable and architecturally attractive space. AD also collaborated with students at the University of Hartford to measure sound transparency and sound absorption of representative wood grate samples, so the impact on the acoustics in Kracum Performance Hall could be assessed. The 400-seat performance hall incorporated the transparent wood grille panels on nearly all visible wall surfaces; the size, shape, and placement of the perforations in the wood, as well as the way the panels are alternately angled within the room have been done in a semi-randomized way to prevent an audible anomaly known as "comb filtering" that could degrade the listener experience.

The research was performed in the Paul S. Veneklasen Research Foundation (PSVRF) Anechoic Chamber at the University of Hartford by Nicholas Roselli, Tyler Cotrell, Paul Mangelsdorf, and Jacob Ott under the direction of Dr. Robert Celmer and Dr. Eoin King. The tests were to measure the acoustic transparency and sound absorption of the wood grilles. The wood grilles were shown to be completely transparent for sound below 5000 Hz, whereas above 5000 Hz, the transparency is reduced by a few decibels. Although some comb filtering was observed, averaging over many sources eliminates the effect. This suggests that in the concert scenario for an ensemble of multiple musicians, this anomaly will not be a concern. The grilles were found to only have a marginal effect on sound absorption. Instead, the wall or curtain behind the grille was the controlling factor.

In addition to the architect and acoustical consultant, the design team included: Schuler Shook (theatre consultant, lighting designer), Acoustic Distinctions, Inc. (audio systems designer, video systems designer), HGA Architects (mechanical engineer), and McGough Companies (general contractor). The background noise level data were measured in the unoccupied space with the HVAC system on.

Small Rehearsal Room | Acoustic Distinctions, Inc.

Concert Hall Closed Panels | Acoustic Distinctions, Inc.

Concert Hall Open Panels | Corey Gaffer

CARLETON COLLEGE

1. DELIVERY 2. GREEN ROOM 3. INSTRUMENT REHEARSAL 4. PERFORMANCE HALL 5. PERFORMANCE PLATFORM

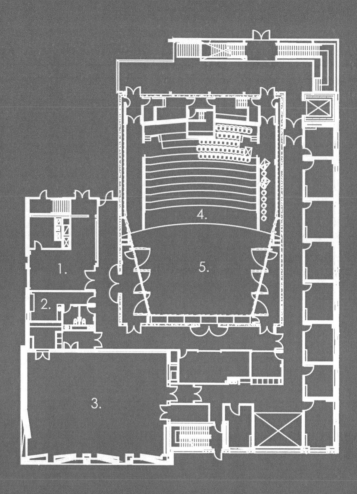

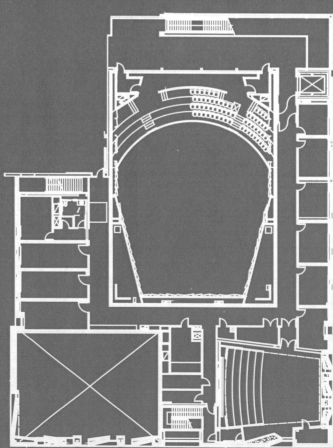

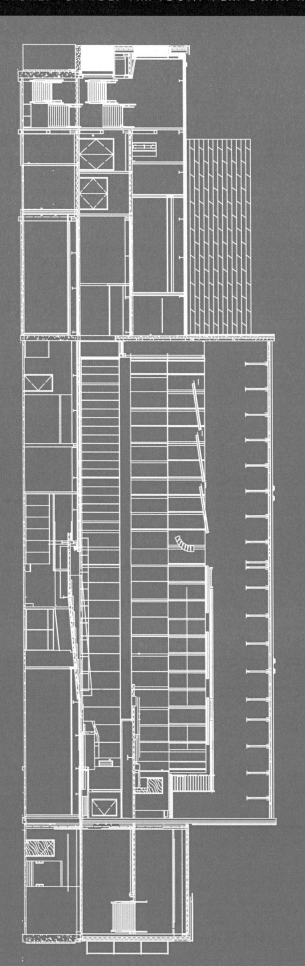

BUILDING SECTION – LONGITUDINAL HALL AND EAST CORRIDOR

ACOUSTICAL CONSULTANT:
ACOUSTIC DISTINCTIONS, INC.

ARCHITECT:
HGA

COMPLETION DATE:
2017

LOCATION:
NORTHFIELD, MN I USA

CONSTRUCTION TYPE:
NEW CONSTRUCTION

CONSTRUCTION/RENOVATION COST:
$28,000,000 (CONSTRUCTION)/
$36,000,000 (TOTAL PROJECT)

FEATURED SPACE DATA:

ROOM VOLUME:
374,640 FT3

FLOOR PLAN AREA:
6,690 FT2

SEATING CAPACITY:
400

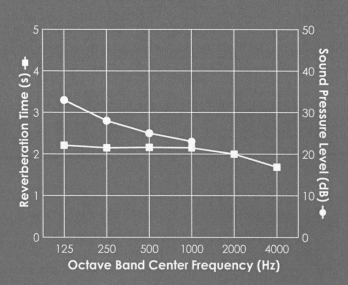

With the construction of this transformative new music building, DePaul University's School of Music is finally given a facility that matches the quality of its faculty and students. The first of this project's three phases is a new 144,000 gross-square-foot music building with the unique 505-seat Mary Patricia Gannon Concert Hall at its heart. Despite a tight footprint and only one balcony, the concert hall is a lofty 64 feet tall. A massive acoustic reflector located 35 feet above the stage covers half the room, providing excellent communication on stage and supportive reflections to the audience for all performances. The open area around the reflector allows the upper volume of the room to be fully energized and fully audible to the audience, while extensive banners in the upper volume and curtains in the lower volume modify the liveness and acoustical character of the room to match each performance.

A soaring central atrium welcomes outsiders, gives students a place to study. and acts as the lobby to the concert hall, two recital halls, and a jazz rehearsal/performance hall. The building also includes large rehearsal rooms for orchestra, band, choir, and percussion. These performance and rehearsal rooms are two or even three stories tall to capture the acoustic volume that is essential to obtain a comfortable balance of loudness and liveness. Percussion studios, chamber music rooms, a recording suite, a dozen classrooms, numerous practice rooms, and underground parking round out this comprehensive facility.

The classrooms, performance halls, and rehearsal

Mary A. Dempsey and Philip H. Corboy Jazz Hall | Kirkegaard

Murray and Michele Allen Recital Hall | Kirkegaard

Choral Rehearsal Room | Kirkegaard

halls are fully equipped with audio video systems, including simple recording systems. The recording suite includes two mix down rooms and a generous live room and is linked to the other performance rooms.

With such an extensive program and limited site, the building had to be tightly packed. Heavy concrete structure, resiliently supported room surfaces, and thoughtful isolation joints maintain excellent isolation throughout the facility.

Phase two, which opened in 2020, was the creation of the 160-seat Sasha and Eugen Jarvis Opera Hall within the shell of the 1950 colonial-style chapel that long served as DePaul's concert hall. The future phase three will be a thorough renovation of the 1960 music building that served as the music school's home and will continue to house the faculty teaching studios and administrative offices.

In addition to the architect and acoustical consultant, the design team included: Schuler Shook (theatre consultant), Kirkegaard (audio systems designer, video systems designer), WMA Consulting Engineers, Ltd. (mechanical engineer), and Bulley & Andrews, LLC (general contractor). The reverberation time and background noise level data were measured in the unoccupied concert hall with half of the banners in the upper volume of the hall exposed and all other banners and curtains stored - a typical setting for orchestra concerts.

Opposite page: Mary Patricia Gannon Concert Hall | Ballogg Photography

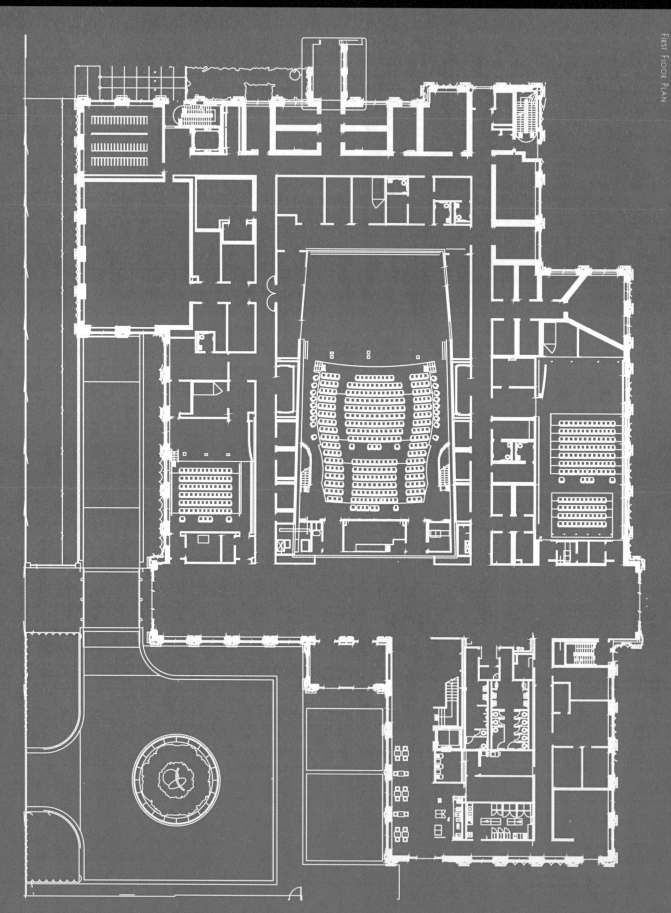

Building Section - North-South

Building Section - East-West

ACOUSTICAL CONSULTANT:
KIRKEGAARD

ARCHITECT:
ANTUNOVICH ASSOCIATES

COMPLETION DATE:
2018

LOCATION:
CHICAGO, IL | USA

CONSTRUCTION TYPE:
NEW CONSTRUCTION, EXPANSION,
ADAPTIVE REUSE

CONSTRUCTION/RENOVATION COST:
$98,000,000

FEATURED SPACE DATA:

ROOM VOLUME:
282,500 FT3

FLOOR PLAN AREA:
8,135 FT2

SEATING CAPACITY:
505

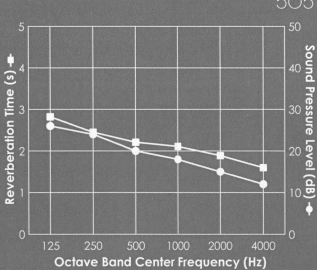

Eastern Connecticut State University (ECSU) Fine Arts Instructional Center (FAIC) in Willimantic, Connecticut is home to ECSU's Performing Arts Department and Visual Arts Department offering 118,000 square feet of instructional, rehearsal, and performance spaces. The jewel of the project is the 400-seat concert hall which serves both the university and the community for a wide variety of events ranging from unamplified instrumental and choral ensembles to amplified events, lectures, presentations, and university functions. The room's retractable curtains and banners located behind the acoustically transparent metal mesh walls allows the acoustics of the room to be customized for each type of event. Warm wood finishes on the lower walls and ample sunlight create a peaceful visual and acoustical setting that is delightful for both audience

and performer. The ceiling reflector is a composite assembly of perforated metal and stretched heavy canvas in an asymmetrical, wave-like design. The canvas provides overhead reflections and allows coupling of the upper volume.

For excellent sound and vibration isolation, the concert hall is surrounded by a structural isolation joint, with only limited connections at major structural beams. Double sets of custom sliding doors allow patrons to flow easily into the performance venue before and after events while providing high isolation during events. A grand glass wall stands at the back of the audience chamber, creating a connection to the outside while maintaining isolation to exterior noise. Heavy floor, ceiling, and wall constructions throughout the building are critical to

CONCERT HALL LOBBY | ROBERT BENSON PHOTOGRAPHY

PRACTICE ROOM | ROBERT BENSON PHOTOGRAPHY

the simultaneous functioning of spaces within the building.

The reverberation time data were measured in the unoccupied concert hall with curtains retracted, and the background noise level data were measured in the unoccupied concert hall with the HVAC system on.

Along with the intricate acoustical design of the concert hall, the hall includes elaborate audio and video systems and connections to support the large range of events. Kirkegaard served as both the audio and video systems designer. The concert hall's A/V system consists of a large video projection with surround-sound audio, sophisticated digital audio mixers, and high-quality main loudspeaker systems. A powerful recording system can capture performances or be used for the mixing and mastering of pre-recorded tracks. Remote-controlled microphone reelers in the concert hall can raise, lower, or position microphones to optimal recording locations. The versatility of the concert hall with regards to both acoustical and audio-visual design allows Eastern Connecticut State University to tailor experiences to a diverse cross-section of specializations, providing an exceptional environment for artists-in-training and the community alike.

In addition to the concert hall, the facility also has two other performance venues: a 250-seat proscenium theatre with flexible seating and a 120-seat studio theater. Theatre Projects Consultants served as the theater consultant for this project. All performance spaces are linked to the extensive backstage support areas including shops, dressing rooms, and offices. The facility also includes a large instrumental rehearsal room and a choral rehearsal room with built-in risers on the ground floor. On the second floor, instructional lab spaces and music faculty offices fill in around the upper volume of the performance and rehearsal venues. The third floor includes art studios, general classrooms and additional offices. In addition to the architects and acoustical consultant, the design team included Altieri Sebor Wieber (mechanical engineer).

Opposite page: Concert Hall | Robert Benson Photography

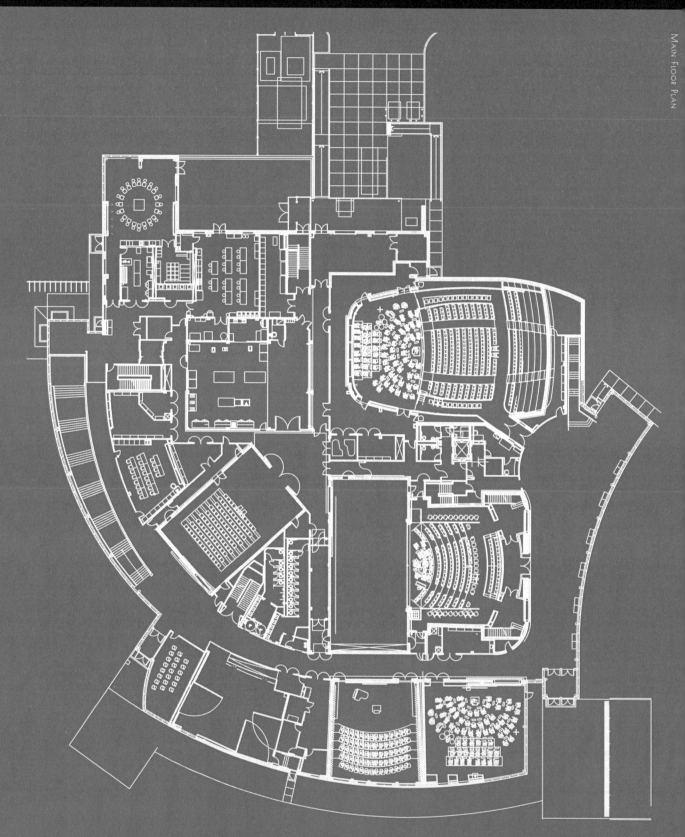

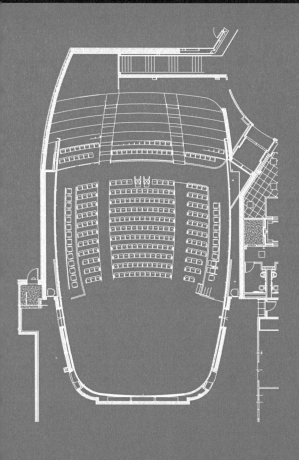

CONCERT HALL – FLOOR PLAN

CONCERT HALL – LONGITUDINAL SECTION

ACOUSTICAL CONSULTANT:
KIRKEGAARD

ARCHITECT:
WILLIAM RAWN ASSOCIATES,
THE S/L/A/M COLLABORATIVE

COMPLETION DATE:
2016

LOCATION:
WILLIMANTIC, CT I USA

CONSTRUCTION TYPE:
NEW CONSTRUCTION

CONSTRUCTION/RENOVATION COST:
$62,000,000

FEATURED SPACE DATA:

ROOM VOLUME:
240,000 FT3

FLOOR PLAN AREA:
5,700 FT2

SEATING CAPACITY:
400

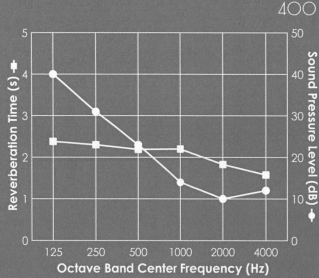

This newly constructed, 64,000 square-foot, state-of-the-art music education building with performance and rehearsal spaces includes an 800-seat concert hall, a 150-seat recital hall, five large rehearsal rooms, twenty individual practice rooms, and twenty-five studios. It also features a recording suite with control and recording rooms.

The 800-seat concert hall incorporates several flexible elements to allow for a very wide range of acoustical environments. Adjustable acoustic curtains can cover almost every wall surface both above and below the lower ceiling elements. The hall also features a large performance platform that can accommodate a full orchestra and choir. For enhanced visual and acoustical intimacy for small-scale presentations, a series of recital towers, which form the upstage wall

at the back of the platform, can be rolled downstage. The concert hall also includes a large Casavant pipe organ that is hidden from view behind a visually opaque yet acoustically transparent wall.

The long section through the hall shows that there is a great deal of additional volume beyond what the viewer sees. The additional volume ensures an appropriately long reverberation time for organ and choral music and also controls excessive loudness for larger and more powerful ensembles. The room volume is typical of a 1,500- to 2,000- seat concert hall even though the seating capacity is 800. As a result of its large and adjustable design, a wide range of reverberation is available to the artists that use the hall. The reverberation time data were measured in the unoccupied space. The background noise level

Concert Hall - Recital Towers | Acoustic Distinctions

Recording Studio | Acoustic Distinctions

Choral Rehearsal Room | Acoustic Distinctions

data were also measured in the unoccupied space with the HVAC system on.

The recording studio control room has been oversized so that it can be used as a teaching environment. There is a window to the large instrumental rehearsal room, which acts as a large recording space, and also a smaller recording studio that can be used for smaller contemporary ensembles (rock, jazz, etc.).

The 125-seat recital hall was designed to be flexible and less formal. There is a large recess in the center of the room with infill platforms including a flat floor space and seating platforms to create a more formal environment for recitals.

The new music building is located only one block from a major train line. In order to ensure that trains are not audible in the performance spaces, a heavy concrete roof was implemented, and the walls were constructed with pre-cast concrete panels.

Hope College is a long-term member of NASM and the new facility design is in compliance with NASM standards for accreditation. In addition to the architect and acoustical consultant, the design team included: Schuler Shook (theatre consultant, lighting designer), Acoustic Distinctions, Inc. (audio systems designer, video systems designer), HGA Architects (mechanical engineer), and GO Construction (general contractor).

Opposite page: Concert Hall - First Level | Acoustic Distinctions

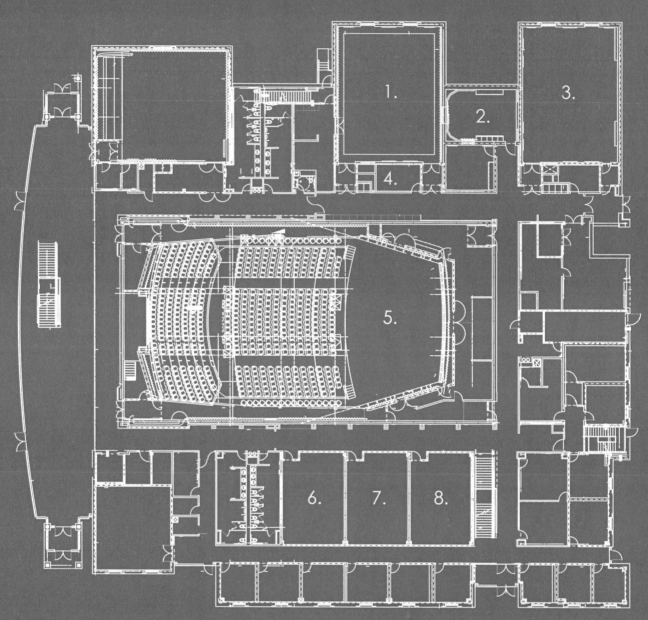

1. ORCHESTRAL REHEARSAL 2. RECORDING FACULTY CONTROL 3. CHORAL REHEARSAL 4. AV RACK 5. ORCHESTRA LEVEL 6. MUSIC EDUCATION 7. KEYBOARD LAB 8. GENERAL MUSIC

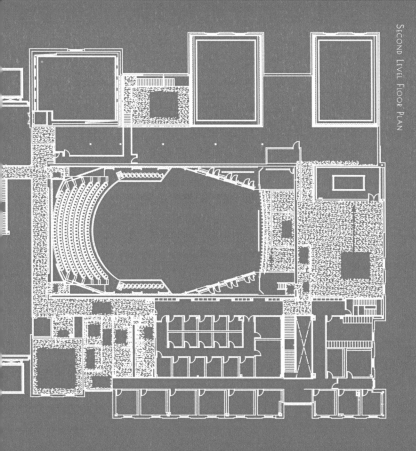

Second Level Floor Plan

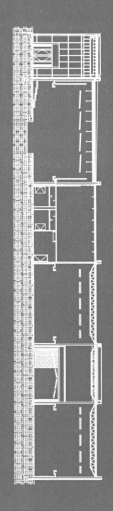

Section - Rehearsal Spaces, Concert Hall, Organ Studio, and Classrooms

ACOUSTICAL CONSULTANT:
ACOUSTIC DISTINCTIONS, INC.

ARCHITECT:
HGA

COMPLETION DATE:
2015

LOCATION:
HOLLAND, MI I USA

CONSTRUCTION TYPE:
NEW CONSTRUCTION

CONSTRUCTION/RENOVATION COST:
$22,000,000

FEATURED SPACE DATA:

ROOM VOLUME:
598,000 FT3

FLOOR PLAN AREA:
9,400 FT2

SEATING CAPACITY:
798

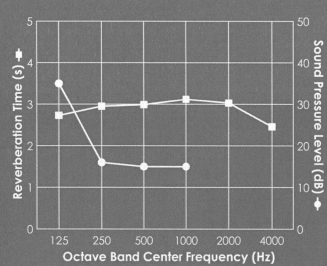

Missouri State University's Ellis Hall was a handsome but aging 1959 modernist music building that had long suffered from poor sound isolation and unremarkable undersized performance spaces. Kirkegaard worked closely with Patterhn Ives on a comprehensive renovation that included replacing the building mechanical systems and much of its curtain wall. Other design team members included: Schuler Shook (theatre consultant, lighting designer), Kirkegaard (audio systems designer, video systems designer), and McClure Engineering (mechanical engineer). The interior of the hall is a virtual gut/rehabilitation, but wherever possible, existing walls were incorporated into new sound-isolating assemblies. The scale of the project compared to the budget required the design team to find carefully-calibrated solutions for what to reuse and what to replace.

The users emphasized the importance of the recital hall, which could not be dramatically improved without increasing volume. The solution was to demolish the hall's floor slab to capture the room volume from an under-used rehearsal room below, replacing the lightly raked audience seating with a more strongly stepped seating zone that greatly improved sight lines, captured more volume, but preserved the original lobby. A handsome wood wall at the side of the hall and a 40-rank Casavant pipe organ at the back of the stage were preserved in place throughout construction, while the rest of the room surfaces were utterly transformed. The result is the gorgeous new 240-seat C Minor Recital Hall.

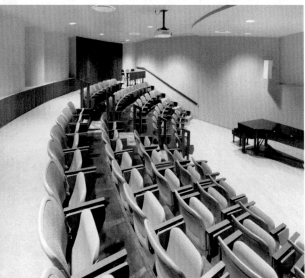

Recital Hall - Sidewall | Architectural Imageworks, LLC

Junior Recital Hall | Architectural Imageworks, LLC

Classroom | Architectural Imageworks, LLC

The reverberation time data were measured in the unoccupied space.

A junior recital hall was created by making modifications to an existing lecture hall with significant improvement to the acoustics for supporting musical performances. The improvements included new upstage wall shaping to provide useful sound reflections from the stage to the audience and rear wall diffusive shaping to break up the undesirable existing concave shape.

In addition to the performance spaces, the building includes twenty-three teaching studios, five classrooms, eighteen practice rooms, and a new lobby space outside the recital hall. Providing the best possible sound isolation between teaching and practice spaces within the limitations of budget and the existing structure was also critical to the success of the project. As part of the renovation, the poorly-insulated original curtain wall at the north and east façades was replaced with a modern insulating-glass curtain wall. In response to the isolation challenge inherent to multi-story glass, the design team laid out the third- and fourth-floor practice rooms and teaching studios as interior spaces that borrowed light from the corridor. Thus, corridors ran along the exterior wall as a buffer. This eliminated the need for costly acoustic mitigation measures at the acoustic weak point inherent to a curtain wall system. Where acoustically sensitive spaces along the exterior curtain wall could not be avoided, secondary inner lites of removable laminated glass were used to protect the rooms from sound carried in the façade. Most of the teaching studios were in a wing with horizontal strip windows. In that area, the acoustical consultant retained any walls that were useful and supplemented them with secondary walls and secondary ceilings to achieve a high level of isolation.

Opposite page: Recital Hall - Stage | Architectural Imageworks, LLC

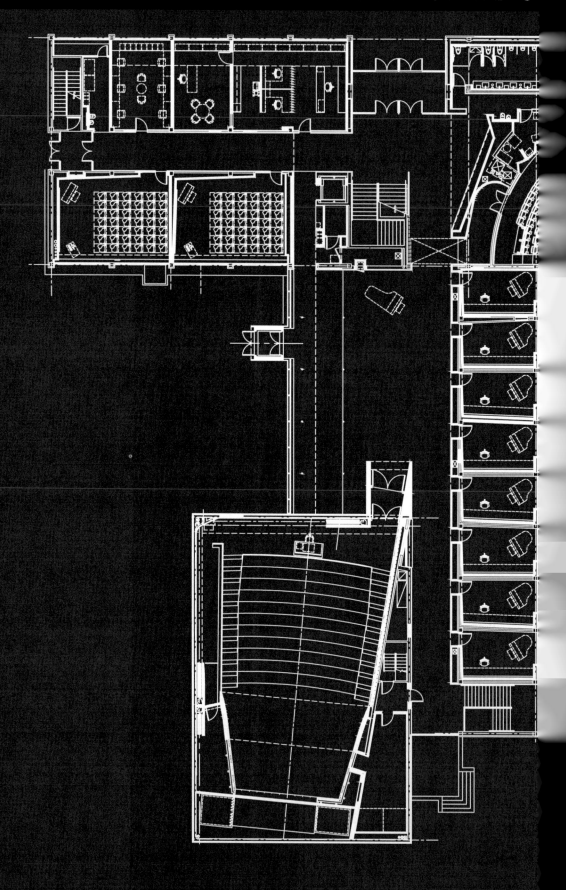

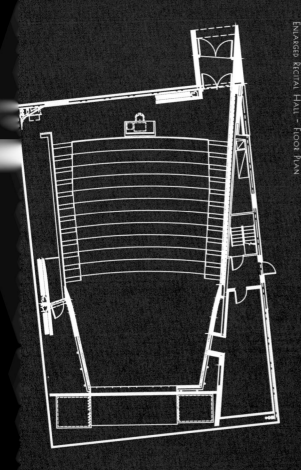

ENLARGED RECITAL HALL – FLOOR PLAN

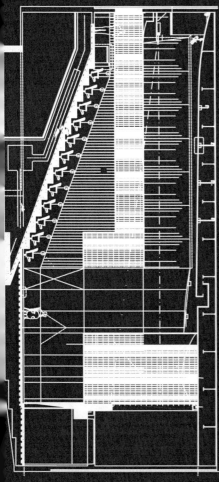

RECITAL HALL – LONGITUDINAL SECTION

ACOUSTICAL CONSULTANT:
KIRKEGAARD

ARCHITECT:
PATTERHN IVES LLC

COMPLETION DATE:
2017

LOCATION:
SPRINGFIELD, MO I USA

CONSTRUCTION TYPE:
RENOVATION

CONSTRUCTION/RENOVATION COST:
$14,300,000

FEATURED SPACE DATA:

ROOM VOLUME:
85,000 FT3

FLOOR PLAN AREA:
3,000 FT2

SEATING CAPACITY:
240

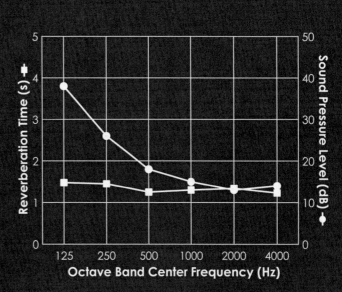

The Nazareth College Arts Center is known for its prestigious music programs and has hosted many prominent performances. The music programs include a diverse range of ensembles, including orchestra, wind ensemble, chamber choir, women's choir, men's choir, jazz ensemble, jazz combos and opera.

In 2016, the college decided to build a new concert hall directly adjacent and attached to their existing music building. A second and related challenge was the project budget of $15.5M, translating to an average construction cost of $580 per square foot for a project that was to include no low-cost spaces. A third challenge was an extremely aggressive project schedule requiring the facility to open in time for the College's 50th anniversary celebrations. All

challenges were overcome thanks to an integrated design process.

The new Jane and Laurence Glazer Music Performance Center at Nazareth College opened to rave reviews in Fall 2018, less than two years after design for this project started. The concert hall, while only seating 550, has a performance platform sized for a full symphony orchestra and choir. In order to acoustically accommodate these large ensembles, the room volume was set to 625,000 cubic feet. Since this size would have required an unusually and impractically tall space, additional volume was captured above some of the hall's circulation areas, backstage, and above a few acoustically sensitive warm-up, teaching, and practice spaces located behind the backstage area. Visually opaque yet acoustically transparent

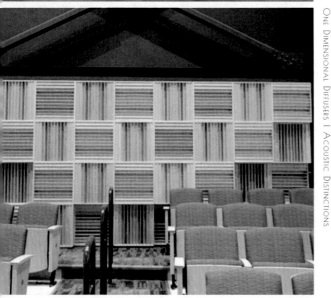

Concert Hall – Rear View | Acoustic Distinctions

Transparent/Visually Opaque Screens | Acoustic Distinctions

One Dimensional Diffusers | Acoustic Distinctions

screens were used to hide the substantial volume that wraps around and above the visual boundaries of the hall's walls and ceiling.

This project had several unique design challenges. For example, it was desired to permit simultaneous use of the concert hall and a warm-up room that sits within the footprint of the upper volume of the concert hall. Substantial, cost-effective sound isolation construction was developed to allow for this goal to be achieved.

This hall has no balcony, leaving a unique, tall rear wall, with no architectural elements to obscure this large surface. The lower section of the wall was covered with one-dimensional diffusers. To increase the effectiveness, the distance from the diffusers to the rear wall was varied; furthermore, some of the diffusers were rotated 90 degrees.

The hall's HVAC systems are virtually silent in operation. The background noise level data were measured in the unoccupied space with the HVAC system on. However, there was negligible difference when the HVAC system was off. In all cases the noise rating fell below RC-15.

Because of the wide range of uses of the hall, there are three levels of motorized, adjustable acoustics curtains. The controller allows the technical staff to simply push a button for various presets depending on the use of the hall. As a result, there is a wide range of adjustability in reverberation available to the hall's users. The reverberation time data were measured in the unoccupied space.

In addition to the architect and acoustical consultant, the design team included: Theatre Projects Consultants (theatre consultant, lighting designer), Acoustic Distinctions, Inc. (audio systems designer, video systems designer), M/E Engineering P.C. (mechanical engineer), and LeChase Construction Services LLC (general contractor).

Opposite page: Concert Hall - Front View | Acoustic Distinctions

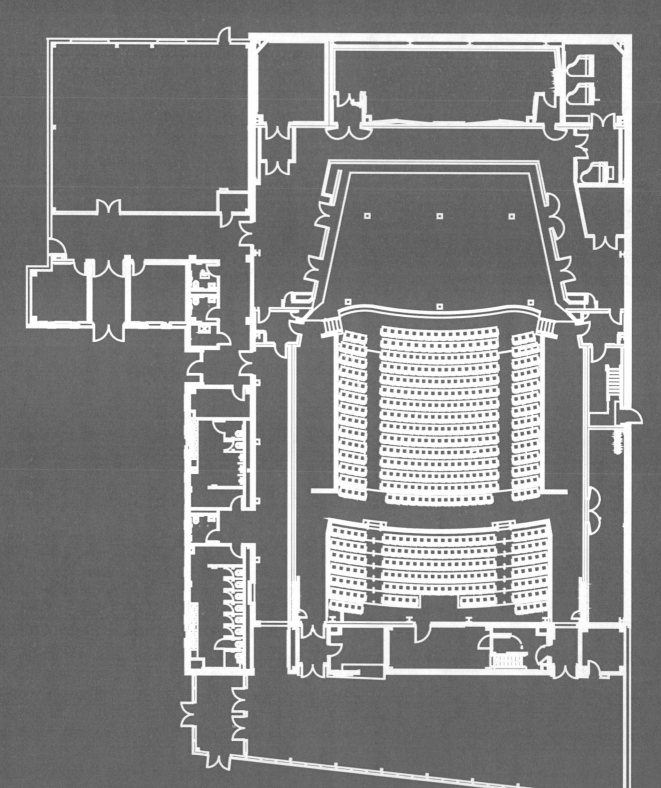

LONGITUDINAL SECTION

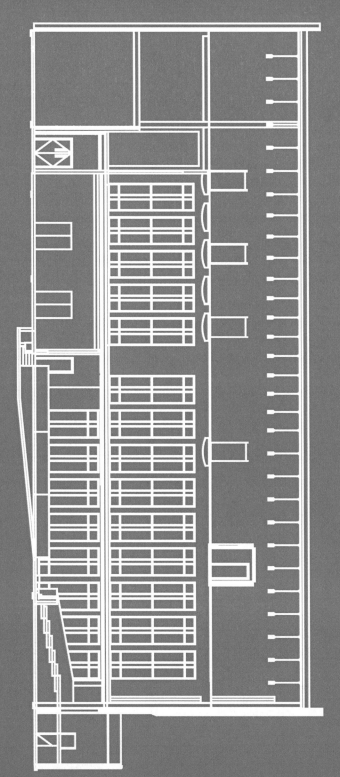

ACOUSTICAL CONSULTANT:
ACOUSTIC DISTINCTIONS, INC.

ARCHITECT:
SWBR ARCHITECTS

COMPLETION DATE:
2018

LOCATION:
ROCHESTER, NY I USA

CONSTRUCTION TYPE:
NEW CONSTRUCTION

CONSTRUCTION/RENOVATION COST:
$15,500,000

FEATURED SPACE DATA:

ROOM VOLUME:
625,000 FT3

FLOOR PLAN AREA:
8,020 FT2

SEATING CAPACITY:
550

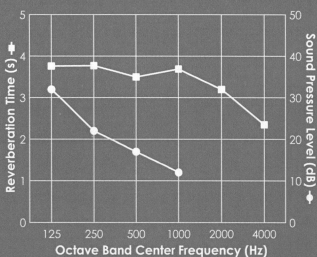

Northwestern's Ryan Center for the Musical Arts is located just south of the university's Pick-Staiger Concert Hall. Programatically, the new building complements the Bienen School of Music's previous home, Regenstein Hall, wrapping the 1970s brutalist building on two sides. This gesture consolidates all of the programs run by the School of Music into one location for the first time since its conception. The expansion also includes the addition of an Arts Green, which connects the Ryan Center, Pick-Staiger, the Block Museum of Art, and the Wirtz Center for the Performing Arts. These connections create a natural stage at the edge of campus for the university's arts community to interact both with Northwestern students and the citizens of Evanston.

A five-story glassy atrium greets users and visitors, boasting magnificent views of Lake Michigan while providing clear direction to the facility's three performance venues. The black box rehearsal/performance space for the opera program also overlooks the lake and features retractable seating and a full tension-wire grid with rigging capability. The choral rehearsal room is similarly flexible, doubling as a junior recital hall with built-in risers that can also serve as raked audience seating.

Most cherished of all is the Mary Galvin Recital Hall, with its warm wood walls bending in and out and running headlong into the upstage wall, which is fully glazed to showcase the Chicago skyline view over the lake. This glass wall presented a concern to both room acoustics and isolation, so the assembly is double-

glazed with laminated lites and a very large airspace, and is angled in section to gently direct sound to the balcony level. A curtain between the lites allows users to control the daylighting without affecting the acoustical support. Banners are incorporated at the undersides of small overhangs created by the wooden ribbons and can be controlled in pairs and groups to customize the hall's reverberation time. Throughout the room, the curvature of the wooden wall elements, as well as the overhead canopies, were carefully studied for diffuse and even coverage for audience members.

The basement uses the entire void beneath the Galvin Recital Hall above as an air distribution plenum. The conditioned air moves slowly to avoid causing noise and drafts at the diffusors beneath the recital hall seats. The displacement ventilation system optimizes comfort and cost of operation. The atrium also uses displacement ventilation, delivering air into a honeycomb floor system that vents around the perimeter. An isolation mat prevents footfall on the stone floor from disturbing the adjacent recital hall.

The reverberation time data were measured in the unoccupied recital hall, and the background noise level data were measured in the unoccupied recital hall with the HVAC and lighting systems on.

The additional program spaces in the Ryan Center include 99 practice rooms, a dozen classrooms that are isolated for sectional rehearsals, a recording suite, administrative offices, music faculty offices, and 35 teaching studios for the piano, jazz, and voice faculty. The glass curtain wall that encloses the majority of the building presented a particular design challenge. Unique details were developed for secondary glazing to achieve a high level of isolation among the practice rooms and teaching studios that share the curtain wall, while still flooding these rooms with natural light.

In addition to the architect and acoustical consultant, the design team included: Schuler Shook (theatre consultant, lighting designer), Kirkegaard (audio systems designer, video systems designer), Cosentini Associates (mechanical engineer), Power Construction Company (general contractor), Thornton Tomasetti (structural engineer), and Hoerr Schaudt Landscape Architects (landscape architect).

Opposite page: Mary B. Galvin Recital Hall | Tom Rossiter

David and Carol McClintock Choral and Recital Room | Tom Rossiter

Shirley Welsh Ryan Opera Theater | Tom Rossiter

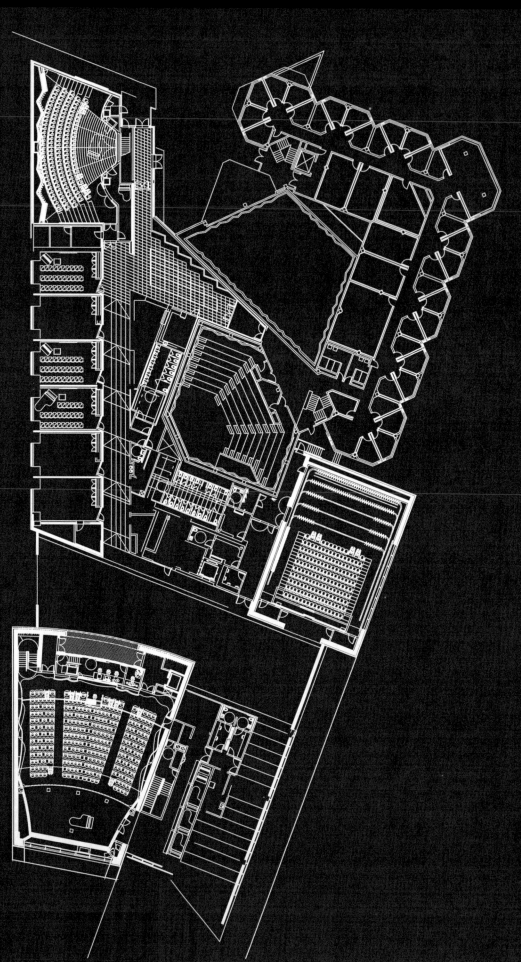

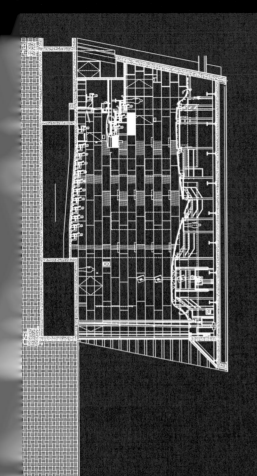

Recital Hall – Longitudinal Section

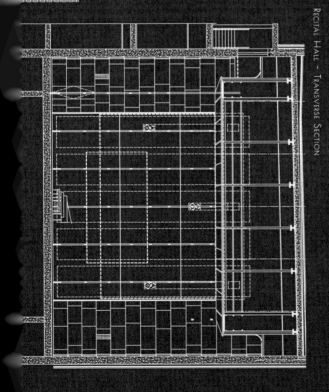

Recital Hall – Transverse Section

ACOUSTICAL CONSULTANT:
KIRKEGAARD

ARCHITECT:
GOETTSCH PARTNERS

COMPLETION DATE:
2015

LOCATION:
EVANSTON, IL I USA

CONSTRUCTION TYPE:
NEW CONSTRUCTION, EXPANSION

CONSTRUCTION/RENOVATION COST:
$108,000,000

FEATURED SPACE DATA:

ROOM VOLUME:
268,800 FT3

FLOOR PLAN AREA:
5,600 FT2

SEATING CAPACITY:
400

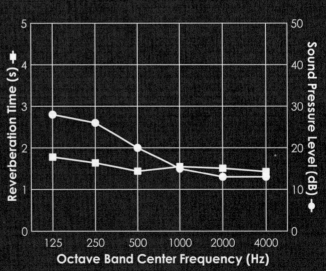

Olympic College is a public community college serving the residents of the Olympic Peninsula of Washington State. The newly constructed College Instruction Center is a 75,000 square-foot building located on Olympic College's Bremerton, Washington campus. The three-story building houses a 275-seat theatre and instructional space for fine arts, theatre, music, and health occupations. It consolidates the instructional resources of four smaller buildings under one roof. In addition to the theatre, the Music Program includes an ensemble rehearsal room, sound stage, practice rooms, percussion practice room, piano lab, and music classrooms.

The ensemble room is a two-story tall room, 40-feet by 35-feet by 28-feet. The interior acoustic design of the rehearsal room is split into two sections, roughly by the "floor" level of the rest of the building. The first section is in the occupied portion of the room, where diffusive lower walls scatter sound throughout the space. In the upper volume of the room, walls reflect sound energy back into the room or absorb energy with deployable velour curtains. When not deployed, the curtains are stowed away in sheetrock garages. The rehearsal room seats up to 75 musicians.

The lower walls are constructed of gypsum board arranged to create a diffuse surface surrounding the musicians. These walls are articulated in a random fashion with long, tetrahedral pyramids running vertically along the side walls. The pyramids are of varied width at their bases and their tips are unevenly spaced along the top of the diffusive wall. The upper

OVERHEAD REFLECTORS | STANTEC CONSULTING SERVICES

ENSEMBLE ROOM - WALL | STANTEC CONSULTING SERVICES

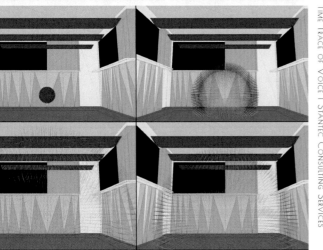

TIME TRACE OF VOICE | STANTEC CONSULTING SERVICES

portion of this lower wall is displaced horizontally by 16 inches, creating a shallow angle to reflect energy back to the musicians. This lower wall section is capped with a horizontal shelf to support the lateral energy distribution.

Overhead, the rehearsal room design includes three fixed reflectors intended to reflect diffuse sound back to the musicians. The ceiling beyond the wooden reflectors incorporates absorption at approximately 40% of its area. A CATT-Acoustic model of the space predicts that 2nd-order reflections begin to permeate the space within 20 milliseconds of the initial sound commencement.

Since the rehearsal room is located directly underneath a classroom, the rehearsal room interior partitions and main ceiling are isolated on a concrete floating floor. At the exterior windows, interior glass is incorporated into the floating room concept and maintains complete separation from all exterior windows.

The reverberation time at 500 Hz in the rehearsal room ranges between 1.10 and 1.65 seconds with curtains fully deployed and curtains stowed away, respectively. Controls for the motorized curtains setting are programmed into a simple control panel placed at the wall for convenient access. The reverberation time data shown were measured in the occupied space. The noise rating shown was calculated for the unoccupied space.

In addition to the architect and acoustical consultant, the design team included: Auerbach Pollock Friedlander (theatre consultant, audio systems designer, video systems designer), Dark Light Design (lighting designer), PAE Engineers (mechanical engineer), Korsmo Construction (general contractor), Magnusson Klemencic Associates (structural engineer), Tres West Engineers (electrical engineer), Coughlin Porter Lundeen (civil engineer), and Nakano Associates (landscape architect).

Opposite page: Ensemble Room | Lara Swimmer Photography

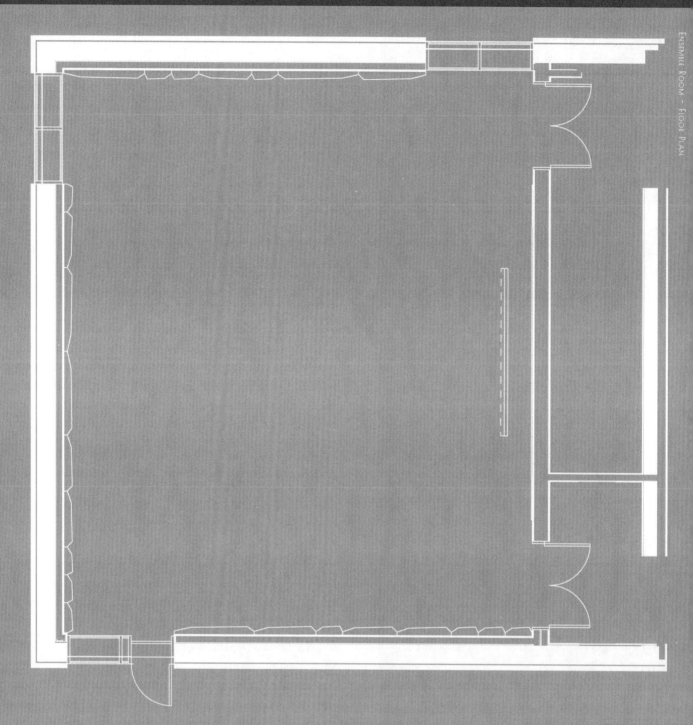

Ensemble Room - Floor Plan

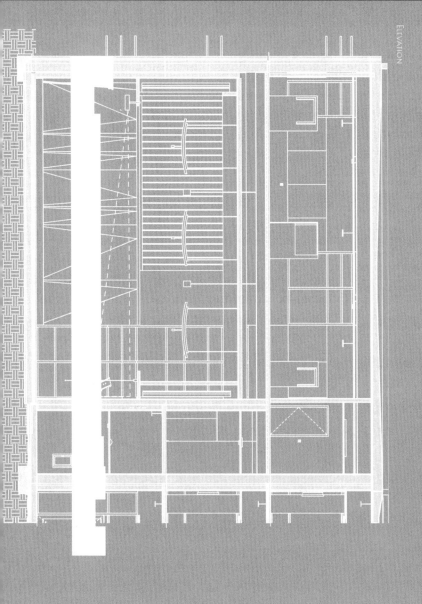

ELEVATION

ACOUSTICAL CONSULTANT:
STANTEC CONSULTING SERVICES

ARCHITECT:
SCHACHT ASLANI ARCHITECTS

COMPLETION DATE:
2018

LOCATION:
BREMERTON, WA | USA

CONSTRUCTION TYPE:
NEW CONSTRUCTION

CONSTRUCTION/RENOVATION COST:
$46,500,000

FEATURED SPACE DATA:

ROOM VOLUME:
40,000 FT3

FLOOR PLAN AREA:
1,425 FT2

SEATING CAPACITY:
35

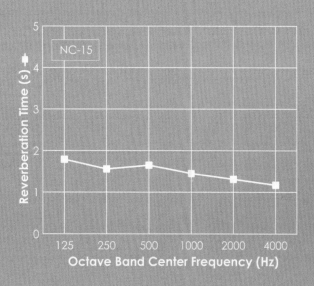

Following the Master Plan for the Arts at Penn State University (PSU), TALASKE developed the acoustic design of a new 14,000-square-foot music building in close collaboration with the architect, theatre consultant, and client. At the earliest stages of the design, the team worked extensively to determine the most desirable configuration for the new recital hall, destined to replace the existing 60's Esber recital hall. Traditional and non-traditional room shapes were evaluated, including symmetric and non-symmetric designs. It is the opinion of this acoustician that most every listening experience in a performance hall is asymmetric in nature, except for listening to an omni-directional instrument at the centerline of the room by a patron also seated at the centerline of the hall. Hence, a hall concept with modest asymmetry was embraced. Furthermore, the

idea of creating a room with seating areas surrounding the stage platform in a vineyard style was considered very desirable, especially where careful observation of performers' technique is commonplace.

The new 400-seat recital hall at Penn State University is believed to be the smallest vineyard style hall designed and constructed to date.

Reverse fan shaping to the rear sections of the audience chamber were employed for the purpose of extending the Early Decay Time of the hall. Sidewalls, working in conjunction with rear walls, created sound reflections which were directed across the room, thereby offering the opportunity to reflect once again off distant side walls and create sound reflections which arrive at patrons well beyond those needed to achieve clarity of sound. This sound reflection pattern contributes

significantly to the *running liveliness* within the hall, a quality of sound deemed essential for the creation of robust reverberation which is audible while music is ongoing. Mid-pitched Early Decay Time values of 1.75 seconds were measured within an approximate one-third occupied hall with all curtains and banners retracted. Mid-pitched Reverberation Times ranging from 1.1 to 2.0 seconds were measured within an approximate one-third occupied hall with all curtains and banners exposed and retracted, respectively.

A significant portion of the walls of the recital hall are precast concrete finished with a skim coat of plaster, directly adhered with no airspace. Remaining surfaces were created with total surface weight in the eight to ten pound per square foot range. These heavy sound reflecting materials contribute to the warmth of sound as heard during acoustic music events.

Windows for exterior views, outward and inward, were strongly desired by the architect. This feature was accommodated acoustically by using thick laminated glass for the inner surface of insulated glazing units and the spanning dimensions of the window units were kept modest to minimize coloration of sound due to panel resonances. The main rear windows are double systems separated by nearly two feet. Inner glazing layers are angled boldly to reflect sound into sound absorbing banners, when deployed, which are oriented perpendicular to windows so views to the exterior are always maintained.

It is the acoustician's opinion that the gross shape of a music hall is best developed using surfaces offering strong, specular sound reflection within a room that is carefully shaped to offer *clarity* of sound, *spaciousness*, and strong *running liveliness*. Once done, some sound diffusion is desirable within the space primarily to diffuse mid- and high-pitched sound and offer a pleasant *timbre*. Within the recital hall, this approach was adopted, and most surfaces offer specular sound reflections. However, via computer modeling, it was found that sound reflections off the upper side walls were somewhat longer delayed, those being greater than 30 milli-seconds longer versus the preceding side wall reflections. As such, it was determined to place the intended sound diffusing finishes on the ceiling to reduce the strength of the upper corner sound reflections within the recital hall.

The recital hall includes an audio system tailored to the room geometry and the acoustic environment. TALASKE was the audio systems designer for the project. A full complement of retractable sound absorbing banners and curtains were integrated into the room, including fully-automated controls. Five sound reflecting towers were designed for use with soloists and small ensembles to offer quickly-arriving sound reflections, avoid long-delayed sound reflections off the upstage wall, and further enhance the very intimate listening experience during performances.

Opposite page: View from House Left Behind Stage | R. Talaske

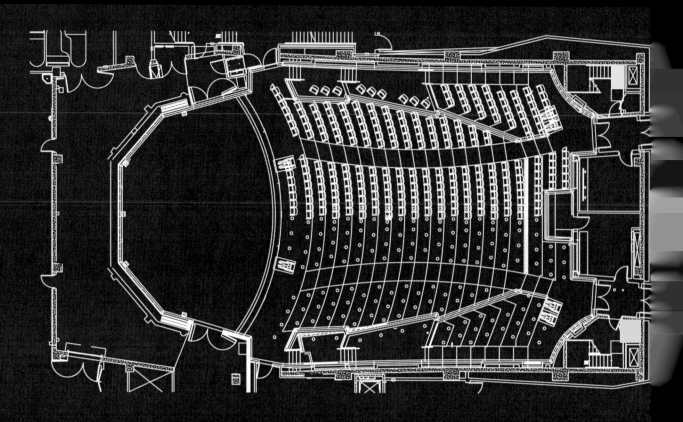

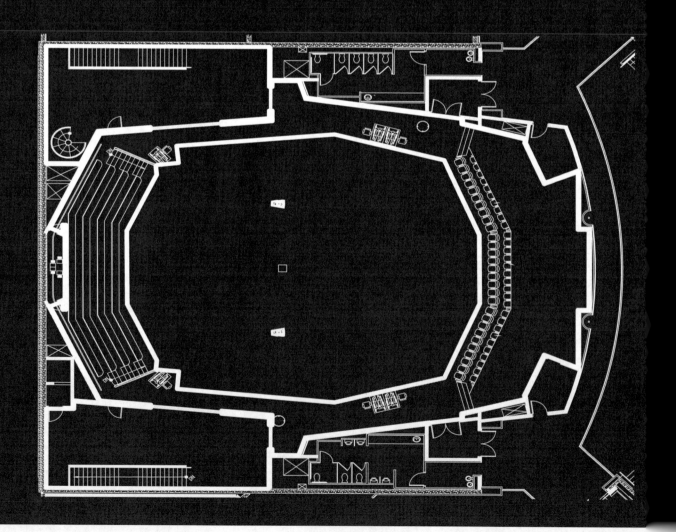

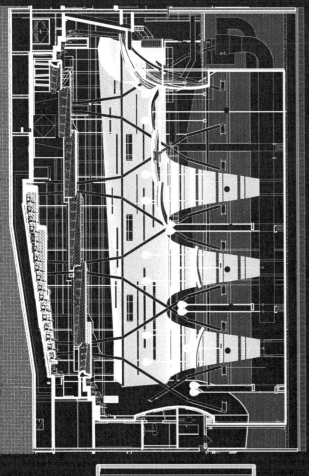

LONGITUDINAL SECTION

CROSS SECTION

ACOUSTICAL CONSULTANT:
TALASKE | SOUND THINKING

ARCHITECT:
BOSTWICK DESIGN PARTNERSHIP WITH
WILLIAM RAWN ASSOCIATES

COMPLETION DATE:
2018

LOCATION:
UNIVERSITY PARK, PA | USA

CONSTRUCTION TYPE:
NEW CONSTRUCTION

CONSTRUCTION/RENOVATION COST:
$26,000,000

FEATURED SPACE DATA:

ROOM VOLUME:
300,000 FT3

FLOOR PLAN AREA:
12,000 FT2

SEATING CAPACITY:
400

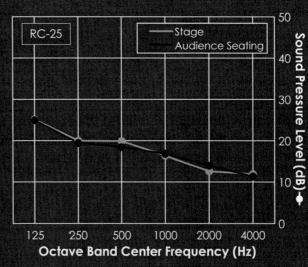

When South Dakota State University (SDSU) originally set out in 1996 to design a new performing arts center for music and theatre, they quickly realized that the available budget was only about half as much as the anticipated construction cost. A decision was made at that time to fulfill the music department's greatest needs (a concert hall) and the theatre department's needs (support space). For many years, the music department continued to use existing space on campus for rehearsal of large ensembles, recitals, teaching studios and practice studios. The theatre department continued to mount their major productions in the existing proscenium theatre.

Fundraising continued after the opening of the Phase 1 facility until sufficient funding was in place to complete the full performing arts center. The Phase 2 addition essentially wraps around the Phase 1 facility. The most significant additions included an 850-seat proscenium theatre, rehearsal rooms for choir and large instrumental ensembles, a 225-seat jewel box recital hall with a pipe organ, music teaching studios, practice rooms, a dance rehearsal room, and two multi-purpose rooms shared by music and theatre.

The recital hall design has its own control room. Both performance spaces have sound light locks with shared star dressing room. The accommodation of the pipe organ into the design of the recital hall was especially challenging because the pipe organ was originally designed and installed in a church that was much larger and differently-shaped. Pipe organ

RECITAL HALL – FRONT VIEW | ACOUSTIC DISTINCTIONS

RECITAL HALL – REAR VIEW | ACOUSTIC DISTINCTIONS

CHORAL REHEARSAL ROOM | ACOUSTIC DISTINCTIONS

INSTRUMENTAL REHEARSAL ROOM | ACOUSTIC DISTINCTIONS

BANNERS | ACOUSTIC DISTINCTIONS

BANNERS RETRACTED | ACOUSTIC DISTINCTIONS

music was not a high priority for the use of the new recital hall; therefore, some special accommodations were made to improve the sound of the pipe organ without compromising the acoustics for the room's primary use as a recital hall.

The considerable room volume of the recital hall, along with its jewel box shape, provide a very responsive yet enveloping acoustical environment. However, the recital hall has a series of retractable curtains that surround the concert platform to provide some flexibility. These curtains are not only extended for larger/louder ensembles, but the curtains on the upstage wall are often extended to improve the clarity of pianos and some instrumental ensembles. Extending these curtains also helps musicians hear more sound return from the hall.

There are also retractable, motorized banners that cover the side walls above the balcony. The primary use of these banners is to compensate for the effect of an audience. Lowering the banners 75% roughly compensates for the difference between an empty house, which is typical during rehearsals, and a full house, which is typical during performances. The pattern in the precast concrete wall panels provide sound diffusion. The panels are rotated to reduce the audibility of comb filtering since the pattern in the precast is so regular.

The adjustable curtains and banners also provide a flexibility in the reverberation in the recital hall. The reverberation time data were measured in the unoccupied space, and the background noise level data were measured in the unoccupied space with the HVAC system on.

In addition to the architect and acoustical consultant, the design team included: Jerit/Boys Inc. (theatre consultant), Acoustic Distinctions, Inc. (audio systems designer, video systems designer), Essential Light Design Studio (lighting designer), ACEI Inc. (mechanical engineer), and Journey Construction (general contractor).

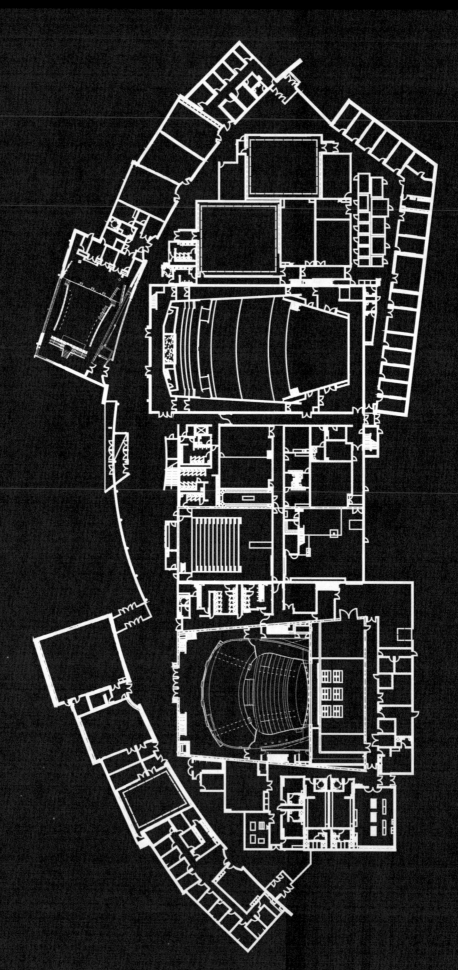

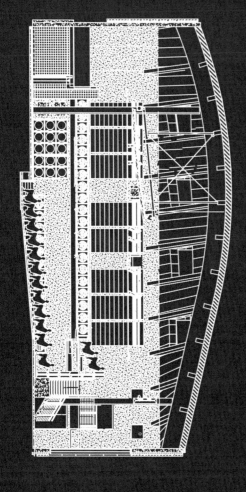

RECITAL HALL - LONGITUDINAL SECTION

TRANSVERSE SECTION AT STAGE

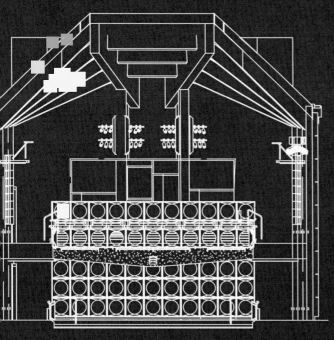

ACOUSTICAL CONSULTANT:
ACOUSTIC DISTINCTIONS, INC.

ARCHITECT:
ARCHITECTURE INC. / HOLZMAN MOSS
BOTTINO ARCHITECTURE

COMPLETION DATE:
2019

LOCATION:
BROOKINGS, SD I USA

CONSTRUCTION TYPE:
NEW CONSTRUCTION

CONSTRUCTION/RENOVATION COST:
$38,800,000

FEATURED SPACE DATA:

ROOM VOLUME:
130,000 FT3

FLOOR PLAN AREA:
3,530 FT2

SEATING CAPACITY:
225

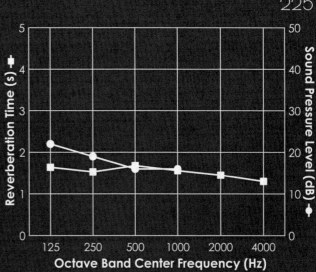

The Buena Vida is a cutting edge "live/learn" facility that houses a growing arts and media program that includes a choral room, a band room with a recording control room, digital editing rooms, piano lab, music practice rooms, lecture hall and an art gallery. This building which serves as an iconic addition to the campus is now realized through the vision of Dr. Craig S. Collins, Dean of the Southeastern University College of Arts and Media.

With the music and visual arts spaces located on the ground floor, each with a multi-layer isolated room-within-a-room design to reduce sound propagation to the upper floors that consist of classrooms, offices and living spaces for students, the five-story building brings the community of talented students and instructors into an active living building which is the "true heart" of the community "in the spirit of Christ." The importance of space planning, sound isolation, volumes of the rehearsal spaces, integration of acoustical and building systems were embedded in the minds of the design team as priorities throughout the design phases of the project. In addition to the architect and acoustical consultant, the design team included: Gramlich & Associates, P.A. (mechanical engineer) and NuJak Companies, Inc. (general contractor). An interactive, collaborative team-oriented approach was used to develop the essential acoustical concepts of sound isolation, sonic structure for music rehearsal and performance and noise and vibration control for building systems.

The Choral Room was built as a stand alone building with the multi-story Buena Vida constructed afterwards around it so that was ultimately included within the new building. The Choral Room set the precedence for the other music rehearsal rooms in the Buena Vida that were specifically and uniquely designed to house the varying ensembles that encompass the School of Music. The Choral Room was designed as a separate building with complete building isolation joints from the raised mechanical room above the Choral storage room. The supply and return air ducts were enclosed within an acoustical shelf that allow early sound reflections to enrich and complement the sound reflections from the curved walls and ceiling panels. The acoustical shelf, while visually complementing the space allowing natural light to reflect onto the ceiling surfaces as well as the room, accommodated the integration of acoustical drapes elegantly and reduced the noise from the air-conditioning system by making extended duct lengths and reduction of duct borne sounds possible.

The 24 ft tall ceiling height in the Choral Room was selected to provide an adequate room volume to naturally enhance the voices of students and sounds of instruments played in the room with a reverberation time of 1.3 to 1.4 seconds that can be reduced to less than 0.8 seconds with the drapes deployed. A network of field-fabricated, custom designed ceiling panels are located to provide cross room reflections to allow vocal ensemble members to

CHORAL ROOM WALLS | SIEBEIN ASSOCIATES, INC.

CHORAL ROOM | SIEBEIN ASSOCIATES, INC.

hear each other and to allow the instructor to hear the students so that music instruction can successfully occur in the room. The ceiling works in combination with the shelf and the curved, site-built wall panels to create a rich, clear sound for the vocalists and their instrumental accompanists in the room. It is a delight to sing, play and listen in this room. The acoustical and architectural design elements used in the Choral Room were used as a system in the other rooms in the building where the balance, locations and amounts of reflecting, diffusing and absorbing surfaces were adjusted to meet the specific acoustical program for each rehearsal room, ensemble room, practice room and teaching studio.

The seamless integration of sound isolation systems with layered acoustical shaping and finishes were carefully designed individually for optimized sound fields for each individual music rehearsal space, large and small. Reverberation times were tuned to each individual space. The reflection sequence in time, amplitude and frequency content or pitch of the essential cross-room reflection that allow students to hear each other and that allow the instructor to hear each student were provided in each room for its several configurations.

From instrumental and vocal ensembles to the individual practice rooms that were acoustically designed to assist the learning student and the professional instructors to further develop their talent and discipline, the spaces provide an atmosphere of encouragement and artistry where the musical abilities of the students can develop and shine. To bring forth optimized volumes of the Band Room, Choral Room and Percussion Studio, each with their own layered composite enclosure, these rooms were located where two-story tall height was possible.

The aesthetically integrated sound diffusing ceiling and wall panels that coexist with the sound absorbent panels and variable acoustic drapes provide the musicians the chance to make the space their own, a canvas that can be colored with the artistry of their music.

The complexities of mechanical, electrical and plumbing systems that supply the physical necessities

for comfort were designed to breathe life into the spaces quietly so that the presence of even the subtlest notes are heard at levels of NC 20 to 22.

The music students and instructors that were formerly located in a typical classroom in a corner of an academic building moved into several temporary trailers with acoustical finishes that were to be relocated to the new facility.

The reverberation time data shown were measured in the unoccupied space. The background noise level data shown were measured in the unoccupied space with the HVAC system on.

CHORAL ROOM WITH DRAPES | SIEBEIN ASSOCIATES, INC.

PERCUSSION ROOM | SIEBEIN ASSOCIATES, INC.

1. Piano Lab 2. Music Library 3. Music Dept. Head Studio 4. Percussion Office Studio 5. Control Room 6. Percussion Practice 7. Choral Rehearsal 8. Band Orchestra Room

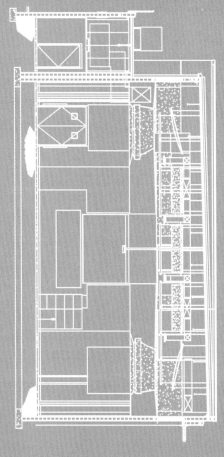

SECTION

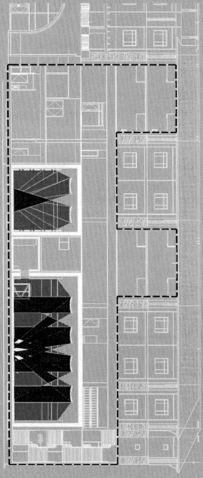

SOUND ISOLATION DIAGRAM

ACOUSTICAL CONSULTANT:
SIEBEIN ASSOCIATES, INC.

ARCHITECT:
KCMH ARCHITECTS

COMPLETION DATE:
2016

LOCATION:
LAKELAND, FL I USA

CONSTRUCTION TYPE:
NEW CONSTRUCTION

CONSTRUCTION/RENOVATION COST:
$25,000,000

FEATURED SPACE DATA:

ROOM VOLUME:
51,353 FT3

FLOOR PLAN AREA:
2,159 FT2

SEATING CAPACITY:
129

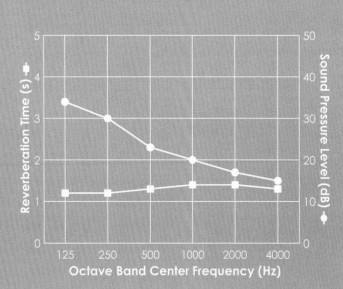

The Ent Center for the Arts is a transformational five-venue center: an innovative collaboration involving the university, six community arts partners, and three local school districts. Jaffe Holden provided acoustic and audio/video systems design services for a wide range of spaces including a multi-use hall, recital hall, art gallery, drama theaters, recording studio, music rehearsal room, dance lab, teaching studios and practice rooms. In addition to the architect and acoustical consultant, the design team included: Auerbach Pollock Friedlander (theatre consultant, lighting designer), ME Engineers (mechanical engineer), JE Dunn Construction (general contractor), Martin/Martin (structural engineer), and NV5 (civil engineer).

The venues are situated one after the other along a lobby corridor. Vertical adjacencies contributed to sound isolation challenges. Corridors were planned to act as buffers between venues but were cut for cost savings. This necessitated the design of complex double hybrid walls of concrete mass and steel studs plus gypsum board and sound control ceilings.

The university president wanted superior acoustics within the 242-seat Chapman Foundations Recital Hall. A generous volume was established for reverberation development. The acoustic consultant designed an innovative, one-piece ceiling featuring slots that provided acoustic breathing room and curved, expansive side walls and areas above the ceiling, rear walls and stage walls. These surfaces are hard and reflective for containment of sound across all frequencies. The stage is a floating-wood

SCHOCKLEY-ZALABAK THEATER | DAVID LAUER PHOTOGRAPHY

STAGE SHELL AND REFLECTORS | JAFFE HOLDEN ACOUSTICS

MUSICIANS ON THEATER STAGE | JAFFE HOLDEN ACOUSTICS

EXTERIOR VIEW | JAFFE HOLDEN ACOUSTICS

construction with a large air space underneath to allow buildup of low frequency resonance. The result is a beautifully resonant room with reverberation times ranging from 2.3 seconds for low frequencies to 1.5 seconds for mid-to-high frequencies (measured in the unoccupied space). This suits a wide range of programming, including choral groups, string quartets, piano, voice, guitar, solo performances, and film screenings. Enhancing this flexibility is an adjustable acoustic curtain system hidden above the partially sound-transparent ceiling and slatted wood wall at the rear of the stage and the upper back of the hall. The background noise level data were measured in the unoccupied hall with the HVAC system on, and the corresponding noise ratings were calculated.

Audio and video systems for the recital hall are unobtrusive yet robust. An acoustically-isolated projector enclosure is recessed into the back of the hall. There is a left and right loudspeaker array, along with front fills and subwoofers flush-mounted in the stage face millwork. The variable acoustics system is comprised of five motorized curtain draw machines and track, plus 2,000 square feet of acoustic drapery and an automation controller.

The Shockley-Zalabak Theater is a 750-seat multi-use hall that supports all types of programming from full orchestra ensembles to small chamber groups. The acoustic response can be adjusted by a motorized system of drapes and acoustic banners concealed above a partially sound-transparent ceiling. A stage lift allows changes in seating configuration and provides a stage extension. An acoustical shell can be configured to accommodate both large and small group performances.

The Dusty Loo Bon Vivant Theatre features fixed acoustics appropriate for natural-voice drama productions as well as amplified musical productions. The space has a low volume and acoustic panels distributed throughout the walls and ceilings. Productions use the full floor in addition to proscenium mode because low background noise allows actors to be heard from any spot in the theater. A black box-style AV system offers flexible system patching and equipment placement, re-deployable to suit any number of audience configurations. More than 20 different connection points connect to the AV equipment to prevent infrastructure from limiting the reconfiguration. Instead of fixed catwalks or a walkable grid, the theater has a rolling gantry catwalk for flexibility and safety. The design ensured that infrastructure did not interfere with use of the gantry. The performance AV systems support high-quality productions and facilitate learning. Student technicians learn and practice the concepts of system operation and signal flow in a professional environment. All the major performance spaces and the ensemble rehearsal room have audio and video tie lines to an acoustically isolated recording studio. Infrastructure and cable passes provide connections to broadcast trucks for broadcast to the university's station or local television stations.

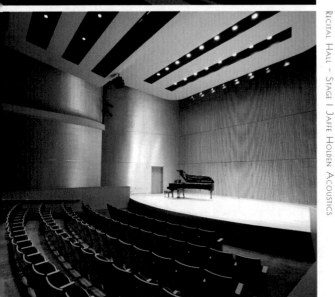

CHAPMAN FOUNDATIONS RECITAL HALL | JAFFE HOLDEN ACOUSTICS

RECITAL HALL – STAGE | JAFFE HOLDEN ACOUSTICS

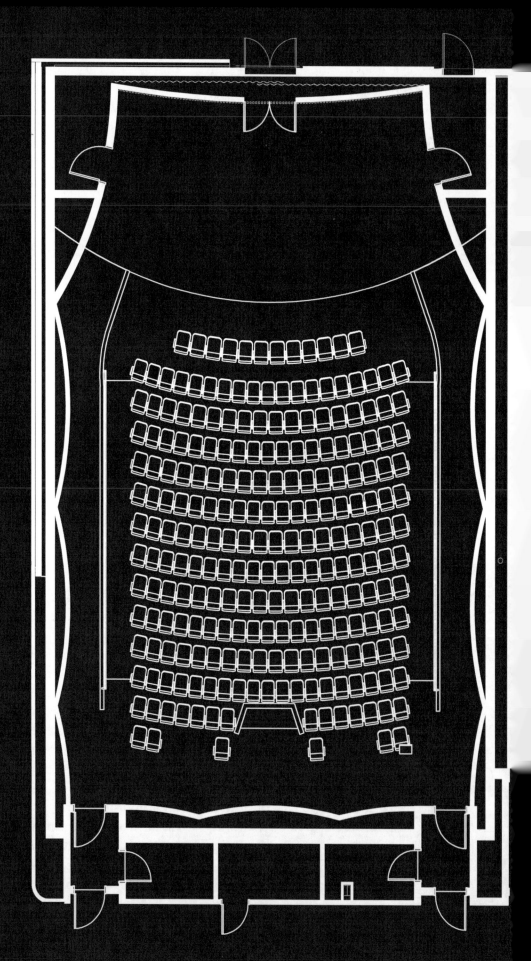

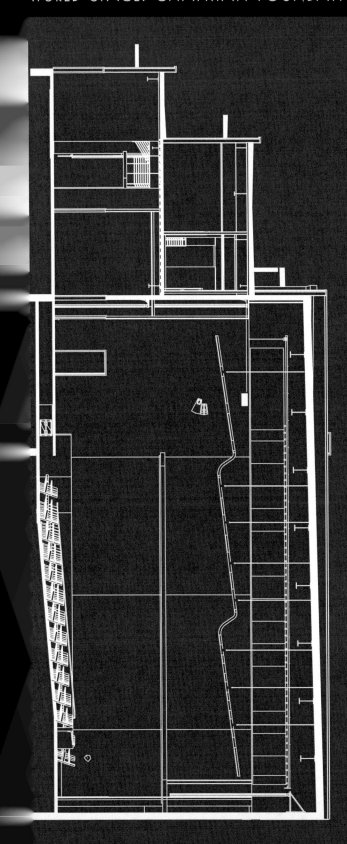

RECITAL HALL – SECTION | JAFFE HOLDEN ACOUSTICS

ACOUSTICAL CONSULTANT:
JAFFE HOLDEN ACOUSTICS

ARCHITECT:
H3 / SEMPLE BROWN

COMPLETION DATE:
2018

LOCATION:
COLORADO SPRINGS, CO | USA

CONSTRUCTION TYPE:
NEW CONSTRUCTION

CONSTRUCTION/RENOVATION COST:
$60,000,000

FEATURED SPACE DATA:

ROOM VOLUME:
132,750 FT3

FLOOR PLAN AREA:
3,300 FT2

SEATING CAPACITY:
242

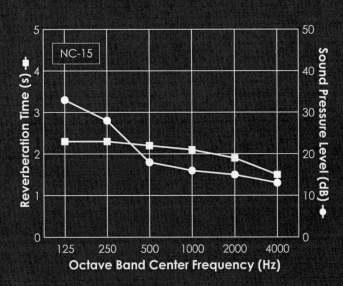

The university's school of music was heavily damaged in 2008 as a result of flooding of the Iowa River. This project involved the relocation and replacement of the school on a site between the campus and the downtown core of Iowa City. The program includes a 700-seat concert hall, a 200-seat recital hall, an organ performance hall, a music library, rehearsal rooms, practice rooms, classrooms, a music production and recording suite and faculty studios and offices.

The busy urban site had a small footprint and a number of acoustical challenges. Space constraints led to vertical stacking of practice rooms, rehearsal spaces and studios. The acoustic consultants utilized highly complex floating concrete floors and box-in-box constructions throughout the building to ensure spaces could be used simultaneously without noisy distractions. The plan called for a performance hall to be located adjacent to a busy street. The acoustical consultant conducted acoustical measurements on site to provide a matrix of acoustic glass and window types to mitigate noise. A window wall was mocked up and tested in an acoustical laboratory to verify viability prior to construction.

The concert hall is the project's prime performance venue, located front-and-center in the six-floor architectural composition. This 700-seat hall is a multi-use space for ensembles, amplified performance and state-of-the-art tracker pipe organ performances. Since the facility's roof could not exceed a certain zoning height, the ceiling was removed in schematic design and replaced with a concrete roof that

Concert Hall – Stage | Jaffe Holden Acoustics

Acoustic Diffusion in Recital Hall | Tim Griffith

Organ Recital Hall | Tim Griffith

Acoustic Kites in Band Room | Tim Griffith

provided mass isolation.

A fixed reflector system was placed below the roof to enhance onstage hearing conditions and provide proper reflections to the audience. The reflector system was extensively modeled to fine-tune the system as it was too costly to have a movable, rigged system. In order to capture the enclosure's entire acoustic volume and not close it off with the reflector, the acoustical consultant modelled the level of sound transparency to the upper volume. This ensured the sound was not too loud nor harsh and that the low frequency sound arriving later from the upper volume added appropriate warmth. The reflector system is a highly complex and intricately-laced suspended theatro-acoustic system. It was designed and coordinated in close collaboration with several

specialists and builders. The main goal of the system is to unify disparate acoustic and theatrical systems within a new, high-performance system. The design criteria—both acoustically and systematically—are complex, interdependent, and often conflicting. Materially, the system is designed to do the extraordinary with the ordinary, using composite aluminum panel in a unique specialty application. From the initial concept stage, a nimble and robust parametric model became the central generative tool, enabling coordination among disciplines to a level of precision never before possible.

Engineers in acoustics, structures, material science, theatrical systems, lighting design, mechanical design, audio/video design and fire protection were looped into a complex orchestration of coordinating and validating progressive design iterations. Specifically, the design team included: Fisher Dachs Associates (theatre consultant), Jaffe Holden Acoustics (audio systems designer, video systems designer), Horton Lees Brogden (lighting designer), Design Engineers (mechanical engineer), and Mortenson (general contractor). As each input was accommodated and synthesized, the digital model was redistributed to the group in a variety of formats for analysis and validation. Construction managers and fabricators gave critical feedback affecting system detailing, componentry, and construction sequencing.

The optimization of systems and drivers was paramount to the initial goal of coordination. The agility of parametric tools further engaged the architects in producing direct-to-fabrication data for construction. The design team used a three-axis CNC (Computer Numerical Control) mill to fabricate prototypes of full-scale components and connections for testing. This unique level of interaction with tools and materials gave critical feedback for system detailing and construction sequencing. Ultimately, the fabrication data for each of the 946 unique panels was generated by this model.

The reverberation time data were measured in the unoccupied concert hall, and the background noise level data were measured in the unoccupied concert hall with the HVAC system on.

Concert Hall Banners – Half Deployed | Jaffe Holden Acoustics Theatro-Acoustic Ceiling | Jaffe Holden Acoustics

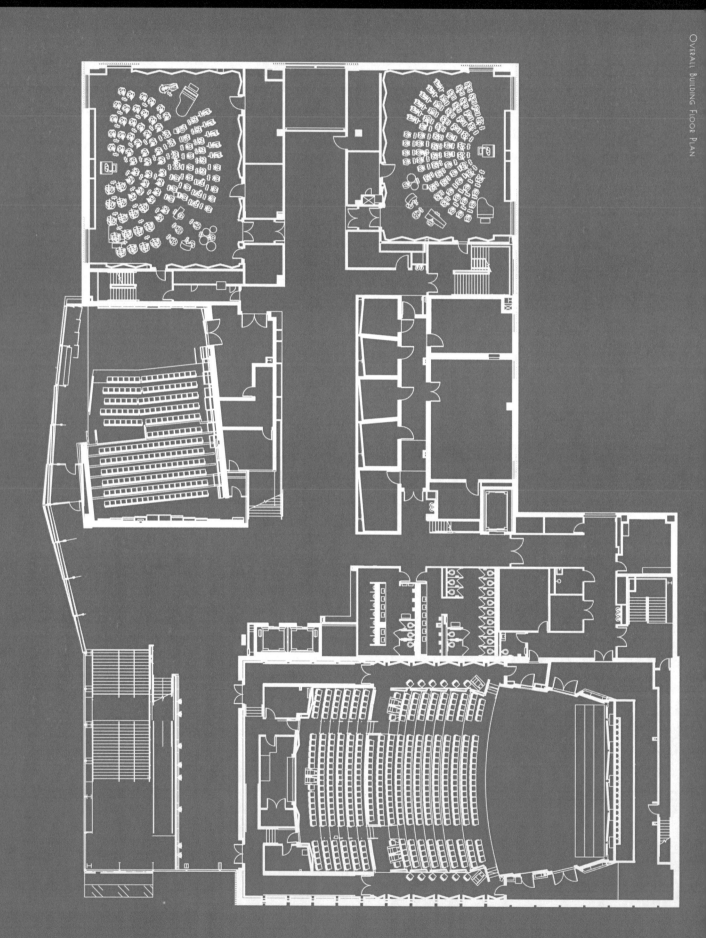

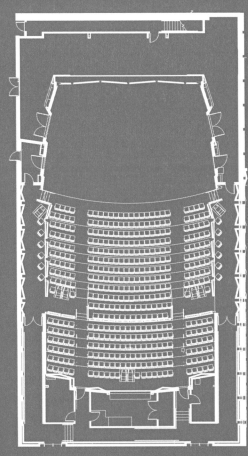

CONCERT HALL - FLOOR PLAN

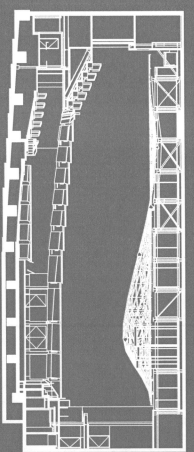

CONCERT HALL - SECTION

ACOUSTICAL CONSULTANT:
JAFFE HOLDEN ACOUSTICS

ARCHITECT:
LMN ARCHITECTS, NEUMANN MONSON ARCHITECTS

COMPLETION DATE:
2016

LOCATION:
IOWA CITY, IA I USA

CONSTRUCTION TYPE:
NEW CONSTRUCTION

CONSTRUCTION/RENOVATION COST:
$112,000,000

FEATURED SPACE DATA:

ROOM VOLUME:
400,000 FT3

FLOOR PLAN AREA:
9,800 FT2

SEATING CAPACITY:
700

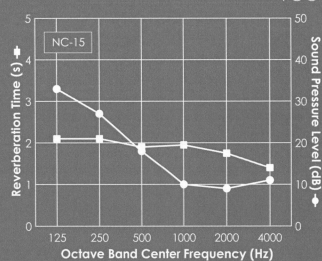

In May 2014 renovation began for the School of Music Swarthout Recital Hall, a 272-seat teaching, rehearsal, and performance facility at the University of Kansas in Lawrence, Kansas. Led by Dr. Robert Walzel, Dean of the School of Music, the Hall (originally constructed in 1957) was completely renovated from aesthetic, theatrical, and acoustical points-of-view. Swarthout is used approximately 300 days a year by students, faculty, and guest performers making it the most used music teaching and performance venue in Kansas. The Hall was reopened on March 30, 2015 with the renovation considered to be a success. The Dean's renovation goal was "it has to take your breath away - both visually and acoustically." The Dean and others believe that this goal was accomplished.

The entire interior of the Hall was replaced down to the stage subfloor excluding some back-of-house spaces. The footprint of the renovated Hall and its volume are basically the same as the original Hall. Following are room acoustic, noise control, and aesthetic goals that were achieved.

1: Hall reverberation time. Prior to renovation the reverberation time at lower frequencies was lower than the mid-frequency reverberation time. This was due primarily to poorly installed 3/4-inch plywood wall panels that produced sound absorption at lower frequencies such that low-register instruments were not well supported. The new wall panels are 1-3/4" thick consisting of gypsum wall board, wood, and wood veneer mounted on heavy steel studs braced to existing masonry block walls. The renovated wall construction, the elimination of a sound absorbing rear wall, and new chairs create a low frequency reverberation time that is higher than the mid-frequency reverberation time, a bass ratio increase from 0.9 to 1.1. The acoustically warmer Hall sound is appreciated when listening to low-register instruments such as cello and double-bass. The mid-frequency reverberation time was maintained at 1.4 seconds partly due to spaces between ceiling panels thus increasing the volume of the renovated Hall. This also provides space for lighting and HVAC supply diffusers. The reverberation time data were measured in the unoccupied space.

2: Uniformity of sound levels throughout the Hall.

Smoothed 1/3 octave-band sound pressure levels for vocal sound from 125 Hz to 8000 Hz at front, center, and rear locations vary by no more than 2 dB and with variance of 3 dB at 100 and 10,000 Hz. Listening confirms that this also exists for instrumental music. Aiding the direct sound are useful sound reflections from wall and ceiling panels with ceiling panels constructed of two layers of 5/8-inch gypsum wall board and walls constructed as indicated previously.

3: Lateral sound reflections. Side walls are oriented to produce lateral sound reflections with a desirable Binaural-Quality Index at mid-frequencies of 0.76.

4: Hall seating. A new wood and concrete floor with a proper rake was installed. Old seating was replaced with fabric-upholstered theatre chairs with wood panels under seats and behind backs.

5: Flutter echo. Previously, there was a significant flutter echo heard about two-thirds of the distance from the stage to the rear wall. This was eliminated by the orientation of new sound reflecting walls.

6: Return of Hall sound to the stage. Before renovation performers described the sound they produced as appearing to "fall off the edge of the stage." Desirable sound reflections are now returned to the stage primarily by the rear wall at suitable levels and timing such that performers are aware that their sound is being transmitted to the Hall seating. The upper portion of the rear wall consists of two layers of 5/8-inch gypsum wall board geometrically shaped to produce scattered sound reflections, to avoid discrete reflections directly back to the stage, and by quadratic residue diffuser panels for the lower portion of the rear wall behind the rear row of theatre chairs.

7: Stage entry. Calipers (ramps) and stairs were installed on both sides of the Hall for easy access to the stage.

8: Piano storage. The lower two panels of the upstage wall can be electrically raised to access storage space large enough to conveniently house two grand pianos.

9: Hall noise. HVAC noise was reduced to NC-26 from NC-45 by replacing all supply duct above the ceiling panels with lined duct, increasing duct size to reduce air velocities, with quiet supply diffusers, and with two duct silencers. Return air noise was reduced by installing grilles in the sides of calipers and using the space under each caliper and behind the adjacent wall as lined return air ducts directing return air into existing return air tunnels beneath the Hall floor. The background noise level data were measured in the unoccupied space with the HVAC system on. A noise control wall was installed between the mechanical equipment room and the Hall. Vestibules were added to reduce noise from the Hall entry corridor. Roof rain noise was significantly reduced by installing mineral-wool insulation and a granular surface on top of the existing 4-inch thick concrete roof and rigid insulation.

10: Sound-video systems. An electrically operated projection screen was installed at the upstage wall with a video projector installed at the center of the lighting bridge. HDMI video inputs and line level audio inputs are available in a stage right floor box and in the control room. The sound system uses a steerable loudspeaker line array on each side of the stage for sound reinforcement when required. These loudspeakers are part of a 5.2 audio-surround system along with a center ceiling panel mounted loudspeaker, two loudspeakers in sidewall panels near the rear of the Hall, and two subbass loudspeakers mounted on each side of the stage behind the stage apron. The 5.2 surround sound system is primarily used for electronic music.

In addition to the architect and acoustical consultant, the design team included: Schuler Shook (theatre consultant, lighting designer), MSM Systems (audio systems designer, video systems designer), Professional Engineering Consulting (mechanical engineer), and Mar-Lan Construction (general contractor). This success was a team effort involving the School of Music Dean; the faculty committee; the architect; the mechanical/electrical engineers; the contractor, their superintendent and construction staff; the theatrical consultant; the KU Endowment Association; contributors to the renovation cost; and the acoustical consultant assisted by architecture and architectural engineering students making this teaching venue renovation also an acoustical teaching experience.

Stage with Ceiling Panels | Robert C. Coffeen

Hall Left Aisle and Caliper | Robert C. Coffeen

Sound Scattering Rear Wall | Robert C. Coffeen

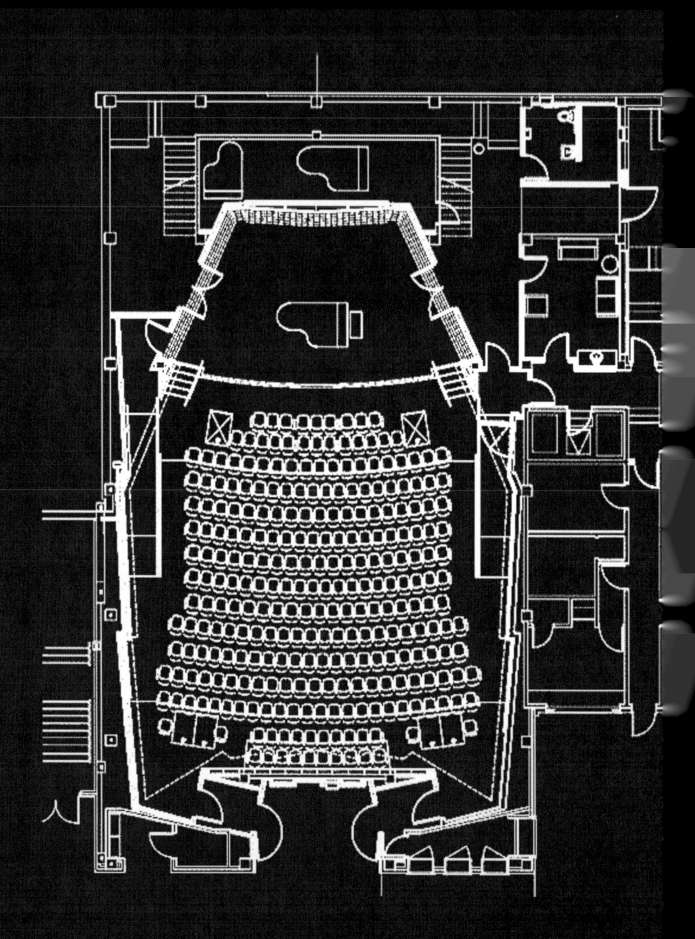

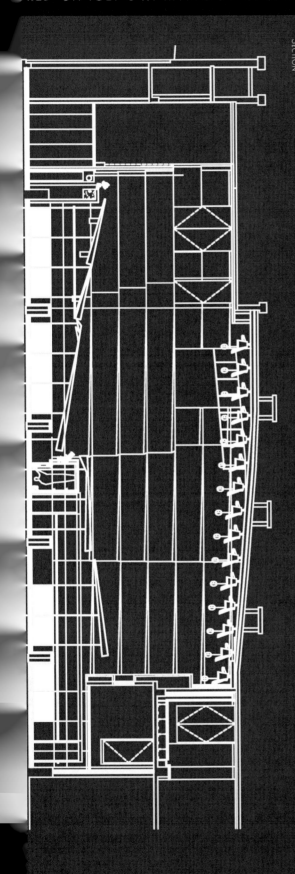

Section

ACOUSTICAL CONSULTANT:
R. C. COFFEEN CONSULTANT IN
ACOUSTICS

ARCHITECT:
SABATINI ARCHITECTS

COMPLETION DATE:
2015

LOCATION:
LAWRENCE, KS | USA

CONSTRUCTION TYPE:
RENOVATION

CONSTRUCTION/RENOVATION COST:
$2,500,000

FEATURED SPACE DATA:

ROOM VOLUME:
81,665 FT3

FLOOR PLAN AREA:
2,800 FT2

SEATING CAPACITY:
272

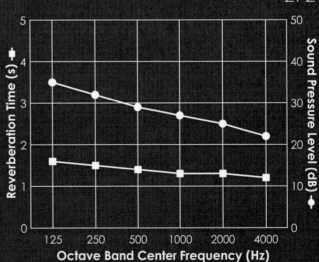

University Hall is the second new academic building in the University of Massachusetts (UMass) Boston Master Plan and at 181,000 square feet, houses a wide range of functions serving several academic departments. Performing arts department spaces include a grand 150-seat music recital hall, a 200-seat flexible theatre, a dance studio, practice rooms, and faculty offices. The building also hosts the art department, chemistry teaching laboratory space, and more than two dozen general purpose classrooms and lecture halls, including a 500-seat lecture hall. The building has a large atrium that serves as a circulation hub for the campus and weaves this interdisciplinary facility together.

The recital hall forms a curved appendage to an otherwise angular building, avoiding any difficult interior adjacencies. The site provided its own challenges, however, as UMass Boston's campus protrudes into Boston Harbor and sits under a major flight path to nearby Logan Airport, an outdoor noise study was conducted and design features were appropriately chosen. A composite concrete roof deck protects the room from aircraft noise above. A portion of the upstage wall is a double window wall system with a very deep, accessible airspace, to provide natural light, views, and excellent sound isolation from outdoor noise.

The hall is essentially a shoebox design with slightly curved sidewalls that taper at the stage. The widest point is 48 feet wide, and the height is 30 feet. The last seating row is a mere 36 feet from the stage, ensuring intimacy. The seating rake is shallow, and two side galleries connect the rear entrance to the stage with no change in floor level. Side and rear walls feature an open wood grillwork up to ten feet in height, and a system of motorized curtains behind the grillwork provides acoustical adjustability. Reverberation times (RT_{mid}, unoccupied) range from 1.5-1.8 seconds, and the perceptual change in clarity is even greater than these numbers suggest. Above 10 feet, the sidewalls are a series of protruding, angled, vertical zigzag strips, that provide high-frequency reflections to the stage and gentle broadband scattering in the vertical dimension. Further scattering is provided by quadratic residue diffuser ceiling panels over the audience. The overall effect is a lush, gentle reverberation surrounding the listener and balancing the strong clarity resulting from the short source-to-receiver distance.

An auralization was created to validate the acoustical design and to demonstrate the utility of the motorized acoustical curtains in adjusting the room for different musical users including vocal choir, piano recital, small orchestra, or jazz band for examples.

The recital hall sound system incorporates loudspeaker technology and electronics to provide both music and speech reinforcement and playback of prerecorded media during lectures. This system maintains distinct sound source localization across three front loudspeaker clusters arranged in a left, center, right configuration. For the main seating area, the cluster point source loudspeakers are larger 3-way horn loaded loudspeakers in 3-box and 2-box arrangements. Coverage patterns are designed to provide even coverage (+/- 3 dB) to every seat, and loudspeakers were chosen to provide excellent directional control and minimize unwanted reflections. The result is realistic sound reinforcement with musical quality and speech clarity. Smaller side fill and front fill loudspeakers provide additional zoned coverage and localization down to the front of the stage. A digital signal processing and control system includes manual and automatic mixer systems along with a fully outfitted multitrack recording system for concerts and student recitals.

The flexible theatre includes a platform floor system to provide flexibility in staging and seating configuration. Two technical gallery levels and a tension wire grid above provide great flexibility in lighting and rigging. Two-inch fiberglass board is affixed to the ceiling and to the underside of the technical galleries to control reverberation. The walls consist of plywood panels at the lower level and gypsum board at upper levels. These sound-reflecting surfaces help to achieve speech intelligibility in thrust or arena formats, in which actors are often not facing audience members. A ceiling of four layers of gypsum board, suspended on spring hangers between steel beams, protects the theatre from potential noise intrusion from chemistry labs above.

Recital Hall – Occupied | Anton Grassi / ESTO Wilson HGA

Black Box Theatre | Anton Grassi / ESTO Wilson HGA

Recital Hall | Anton Grassi / ESTO Wilson HGA

The theatre sound system provides sound reinforcement during theater events and provides students an opportunity to gain experience in system configuration for audio in theater productions. A variety of portable loudspeakers, microphones, and electronic playback equipment can be connected through a flexible patching system. These components are configured on a show-by-show basis to provide sound reinforcement, theatrical sound effects, and playback of prerecorded audio. Semi-permanent loudspeaker clusters are installed above the tension wire grid to maintain sight lines and easy setup for daily use during set construction, rehearsals, and University events. Additional sound system equipment includes technician headsets and backstage paging. The flexibility of this system and its components are designed to meet the needs of current and future theater sound designers.

Music practice studios are located on the second floor and are modular rooms by Wenger, each with independent floating floors. Other acoustical features in the building include a fourth-floor dance studio on a floating concrete slab above a classroom, and upgraded glazing in many spaces – informed by environmental noise measurements of aircraft flyovers from the roof of an adjacent building.

In addition to the architect and acoustical consultant, the design team included: Theatre Projects Consultants (theatre consultant), Cavanaugh Tocci Associates, Inc. (audio systems designer, video systems designer), BR+A Consulting Engineers (mechanical engineer), Gilbane Building Co. (general contractor), and LeMessurier (structural engineer). The reverberation time data and noise ratings were calculated for the unoccupied space.

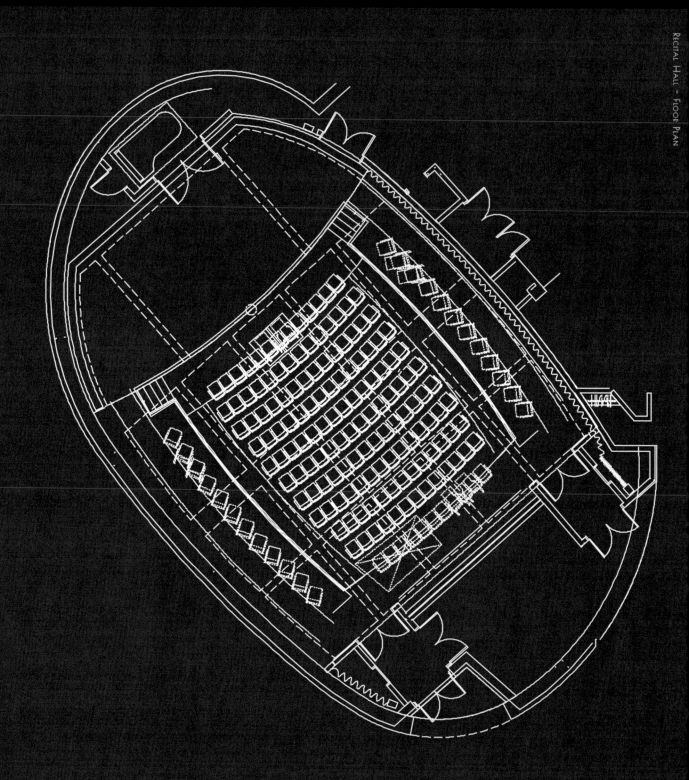

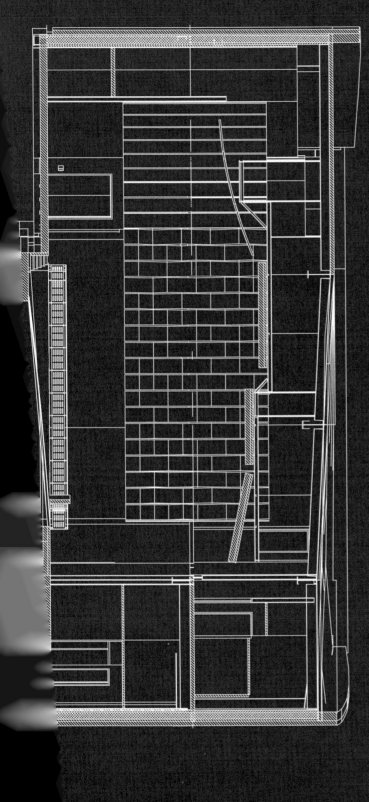

RECITAL HALL - ELEVATION

ACOUSTICAL CONSULTANT:
ACENTECH INCORPORATED

ARCHITECT:
WILSON HGA

COMPLETION DATE:
2016

LOCATION:
BOSTON, MA | USA

CONSTRUCTION TYPE:
NEW CONSTRUCTION

CONSTRUCTION/RENOVATION COST:
$130,000,000

FEATURED SPACE DATA:

ROOM VOLUME:
88,000 FT3

FLOOR PLAN AREA:
3,028 FT2

SEATING CAPACITY:
150

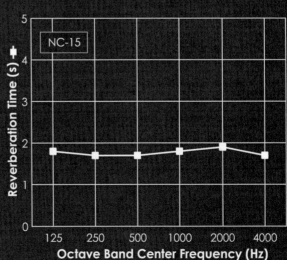

The Willis S. and Janet A. Strauss Performing Arts Center (SPAC) on the University of Nebraska Omaha (UNO) campus has been home for the School of Music since its construction in 1971. The School is dedicated to all aspects of music education and performance but found that limitations with the existing building required renovation and expansion to support a growing program. The building is now comprised of two components: the partially renovated 1971 building and the 2018 tower addition.

The original building contains the 460-seat concert hall with its Casavant Frères Opus 3603 Organ, large instrumental and vocal rehearsal rooms, 27 small and medium sized practice rooms, a green room, director's office and administrative suite, teaching offices, classrooms, and lounge areas. The 2018 renovation focused on design elements deemed crucial for core use with renovations primarily focused on classrooms, public areas, and the concert hall. Little was done at existing practice rooms with the exception of enhancing sound isolation by sealing penetrations made over the building's 45+ year history, lagging major flanking paths, and upgrading door seals. Future goals include further upgrades to the concert hall stage surround, large rehearsal rooms, and practice room finishes.

The 200,000-cubic-foot concert hall has been renovated to improve communication on stage, optimize frequency balance, add variable acoustic control, and quiet background noise. Hinged panel wood/carpet variable absorption at rear walls was replaced with a vertically rising banner system. Additional banners were added at side-wall areas previously without variable control. These new elements supplement existing curtains located behind a wood grillage at upper side walls. All new banners store in robust housings high in the space to limit their impact when retracted. A fine-scaled diffusive finish was added to rear wall areas to control previously harsh reflections that previously required exposure of the carpet most of the time.

The original stage canopy did not sufficiently support the entire stage, so original canopy panels were replaced with new GFRG elements with a more optimal shape for both on-stage communication and projection to the audience. New GFRG panels were added in the upstage area to further aid stage communication while not obstructing the organ. A new performance audio system and video projection were added as part of this renovation. Background noise in the existing concert hall was a major concern prior to the 2018 renovation due to an extremely short path between the mechanical room and concert hall. Custom attenuation strategies, along with a circuitous architectural plenum within the concert hall dramatically reduced the background noise by over 20 dB below 500Hz.

The 4-story tower, added in 2018, provides much needed program space, including a 100-seat recital hall, 1,600-square-foot recording suite, piano

SOUTHEAST EXTERIOR RENDERING | HDR ARCHITECTURE, INC.

RECITAL HALL RENDERING | HDR ARCHITECTURE, INC.

studio, collaborative piano office, 860-square-foot percussion rehearsal space, percussion studios and practice rooms, teaching offices, lounge spaces, and storage.

The new recording suite is focused on teaching students the art of recording and mastering in an environment suitable for such delicate work. The recording suite includes a 700-square-foot tracking room connected to a 640-square-foot recording control room. Robust window and wall systems provide required sound isolation. The original concert hall and most rooms in the addition are connected to the recording control room for archival and editing capability. Within the recording suite, smaller 130-square-foot and 80-square-foot isolation rooms gracefully handle anything from drum set tracking to individual instrument and voice recording. The suite is sandwiched between the recital hall and percussion floor of the tower, so barrier ceilings and robust flooring were provided to allow for simultaneous use.

The core of the new addition is a smaller stage needed to host recitals and smaller performance groups. The new 4,700-cubic-foot recital hall has a stage 36-feet at its widest and 21-feet deep, seating up to 100 people on a flat floor with a raised stage. In addition to performance, this room is designed as a rehearsal and classroom space, accommodating smaller string and wind ensembles, choir, opera, masterclasses, and workshops.

Macro wall shaping in the recital hall provides early reflection support and diffusion. Full-frequency reverberation is supported by robust masonry walls, which structurally "float" within the steel structure of the building. Fine-scale diffusion is incorporated into the finish masonry, progressing from smooth, ground-face CMU (concrete masonry unit) at the stage to split-face and offset CMU at the rear. Portions of the CMU are intentionally left unsealed to take advantage of the natural porosity of the material, subtly tailoring the high-frequency reverberation in the space, yielding a warmer, well balanced sound. Reverberation and early reflections are controlled with vertically deployed sound absorptive banners. The reverberation time and background noise level data shown were calculated for the unoccupied recital hall. AV systems include performance audio for jazz and rock groups as well as speech and video presentations and easily deployable video projection.

Visual connection outside the building is a key architectural feature of the new addition. This presented acoustic challenges, particularly for the new recital hall and sensitive recording areas due to the building being situated between U.S. Highway 6 (Dodge Street) and a 47-bell campanile, a centerpiece of the UNO campus. Robust glazing systems and isolation wall construction provide views yet assure that required low background noise levels are satisfied in all areas.

In addition to the architect and acoustical consultant, the design team included: Threshold Acoustics LLC (audio systems designer, video systems designer), HDR Architecture, Inc. (lighting designer, mechanical engineer), and The Weitz Company, LLC (general contractor).

Percussion Studio Rendering | HDR Architecture, Inc.

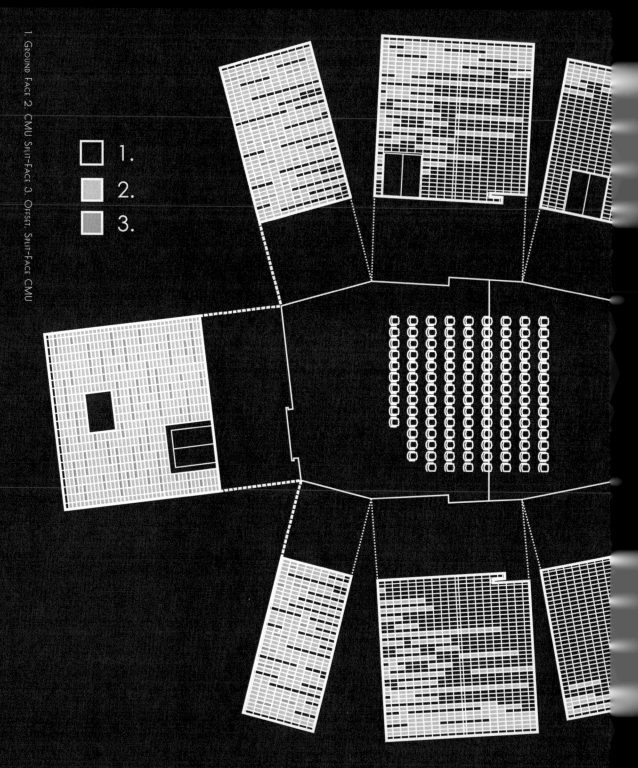

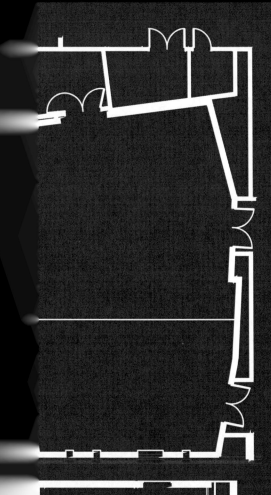

Recital Hall – Floor Plan

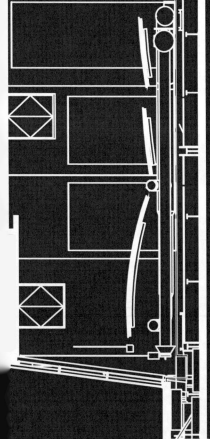

Recital Hall – Section

ACOUSTICAL CONSULTANT:
THRESHOLD ACOUSTICS LLC

ARCHITECT:
HDR ARCHITECTURE, INC.

COMPLETION DATE:
2019

LOCATION:
OMAHA, NE | USA

CONSTRUCTION TYPE:
NEW CONSTRUCTION, RENOVATION

CONSTRUCTION/RENOVATION COST:
$18,000,000

FEATURED SPACE DATA:

ROOM VOLUME:
47,000 FT3

FLOOR PLAN AREA:
1,800 FT2

SEATING CAPACITY:
100

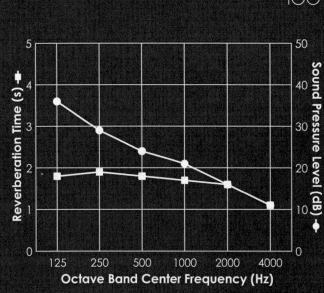

The venerable Oregon Bach Festival (OBF) has long presented concerts of Bach, his contemporaries, and modern music that draws inspiration from early and baroque music. For years, OBF rehearsed and performed at various venues around Eugene, relying heavily on The University of Oregon's (UO) Frohnmayer Music Building and Beall Recital Hall. Since 2015, the Berwick Academy for Historic Performance has complemented the Festival with an intensive tuition-free training course capped by performances in the Festival. At last, OBF has a home of its own, nestled next to UO's music building.

In addition to the architect and acoustical consultant, the design team included: The Shalleck Collaborative, Inc (theatre consultant, audio systems designer, video systems designer), O-LLC (lighting designer), Glumac (mechanical engineer), Madden & Baughman Engineering, Inc. (structural engineer), Cameron McCarthy (landscape architect), and Capital Engineering & Consulting (civil engineer).

Berwick Hall includes office and meeting space for OBF, with the boardroom and one of the offices suitable for auditioning musicians. The heart of the building is a spectacular rehearsal room, designed specifically for early music. Soloists, small ensembles, choirs, and chamber orchestras can rehearse, perform, and record in the room, and students can hear lectures and interact directly with musicians who are experts in early music.

Two walls of the rehearsal room are wood, shaped in sweeping curves that gradually direct sustained sound upwards. Two walls are split-face block,

BERWICK HALL INTERIOR | KIRKEGAARD

FROHNMAYER MUSIC BUILDING LARGE REHEARSAL

BERWICK HALL EXTERIOR | KIRKEGAARD

MUSIC BUILDING CLASSROOM

finished with limewash as a wall would have been in Bach's day. These walls scatter high frequencies and gently absorb some of the very highest frequencies, for a warm, smooth sound. The sweeping ceiling high overhead provides easy communication and beautiful reverberance. Skylights and clerestory windows are well isolated to preserve a quiet ambience. Motorized banners allow the musicians to adjust the balance of reverberation and clarity for each rehearsal. These acoustic banners have five preset positions. Preset 1 corresponds to fully stored; whereas presets 2-5 correspond to the banners being deployed in 5-foot increments. The unoccupied mid-frequency reverberation time ranges from 1.3 seconds with the banners fully deployed to 2.5 seconds with the banners fully stored. The reverberation time data was measured with 10 feet of the banners exposed in the room. Virtually silent air conditioning keeps everyone comfortable while ensuring that every nuance of a performance can be heard. The result is a remarkable, one-of-a-kind room. The background noise level data were measured in the unoccupied space with the cooling systems on, and some noise was audible from a unit outside of the room through the building structure.

In 2008, Kirkegaard worked with Bora (architect) on the design of two additions to the Frohnmayer Music Building totaling 30,000 square feet of spaces to complement and tie together the loose ends of the existing 1920s, '50s, and '70s spaces. One addition includes three classrooms of various sizes, specifically equipped and designed for music instruction and sectional rehearsals, and a three-story tower containing a student lounge, practice rooms, and well-isolated, comfortably-sized teaching studios. The other addition accommodates the jazz and percussion programs in a one-story wing. The large jazz rehearsal room in the center of the wing is designed to work as a live room for jazz recordings. An even larger ensemble rehearsal room terminates the wing. This tall, handsome room is filled with natural light and has extensive adjustable curtaining to accommodate everything from an orchestra rehearsal to a percussion ensemble to a lightly-staged opera performance.

Careful shepherding of a tight budget allowed the new construction to be built at a high-quality without extravagance, freeing up funds for light renovations of many of the existing spaces. The design included innovative use of HVAC systems, such as displacement ventilation at the ensemble rehearsal room, radiant floor, and passive thermal mass to create an unusually "green" music facility.

FROHNMAYER MUSIC BUILDING LARGE REHEARSAL – CANVAS REFLECTORS

FROHNMAYER MUSIC BUILDING LARGE REHEARSAL – DETAIL

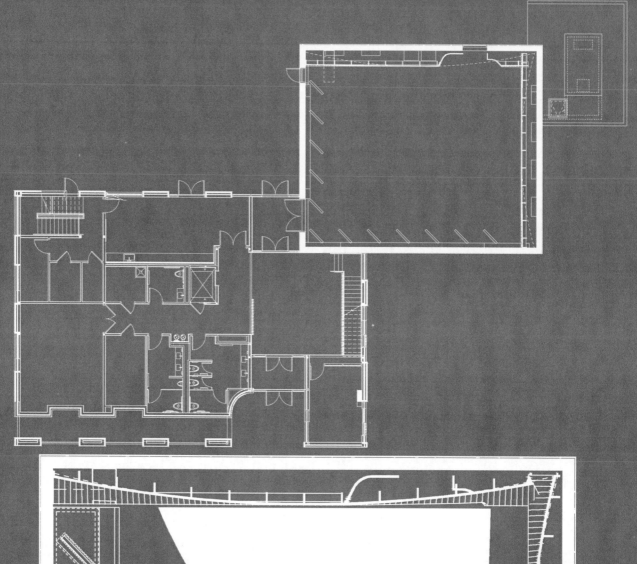

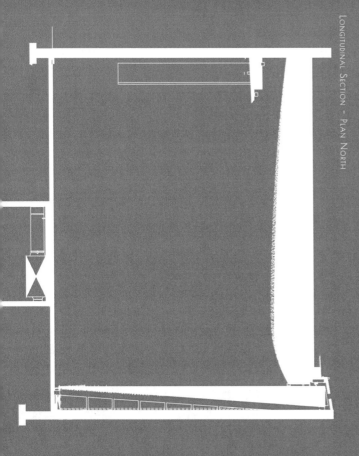

LONGITUDINAL SECTION - PLAN NORTH

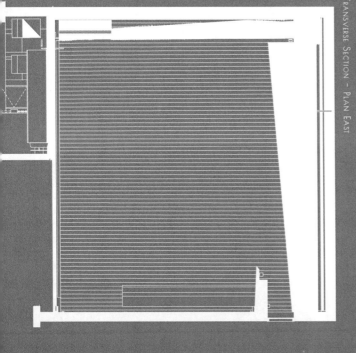

TRANSVERSE SECTION - PLAN EAST

ACOUSTICAL CONSULTANT:
KIRKEGAARD

ARCHITECT:
HACKER ARCHITECTS

COMPLETION DATE:
2017

LOCATION:
EUGENE, OR I USA

CONSTRUCTION TYPE:
NEW CONSTRUCTION

CONSTRUCTION/RENOVATION COST:
$6,900,000

FEATURED SPACE DATA:

ROOM VOLUME:
64,300 FT3

FLOOR PLAN AREA:
2,000 FT2

SEATING CAPACITY:
140

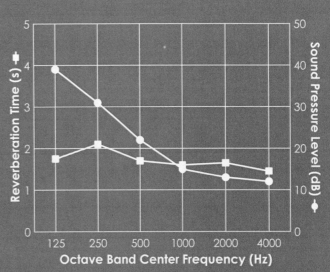

hitworth University's Cowles Music Center is a duet between musical art and architectural form that successfully exhibits the functionality required in a university music structure. They each share design, rhythm, tempo, and scale. The university desired to elevate the campus's visual and musical experience while creating a modernized facility for exemplary instruction, anticipated program growth, and increased community outreach. Originally built in 1978, the university's high-profile 25,100 square-foot, two-story, mid-century music building underwent 15,625 square feet of upgrades and a 21,415 square-foot new addition to achieve these objectives. The design team thoughtfully paid homage to the existing building's architectural fabric and the overall campus context to create a more modern, engaging, and timeless structure.

The west elevation is a composition of glass and ground-face block to create an intentional lower scaled mass and front door overhang. Inviting and transparent entry statements at each end increase front door exposure from both east and west and allow the eye to fill the gap between two eras of architecture.

The chosen, linear relationship of the addition to the existing building was leveraged to form an open, inner circulation spine as a transitional element to bridge old to new. This two-story, day-lit "Main Street" runs the entire east-to-west length of the building. This "gasket" delicately allows the new addition to functionally and architecturally touch the former north exterior elevation creating a classic

CHORAL REHEARSAL ROOM | LARA SWIMMER PHOTOGRAPHY

INSTRUMENT REHEARSAL ROOM | LARA SWIMMER PHOTOGRAPHY

MULTI-USE ROOM | LARA SWIMMER PHOTOGRAPHY

STUDENT LOUNGE | LARA SWIMMER PHOTOGRAPHY

brick backdrop for all open, interior common areas. Hardness of exposed structural steel and metal decking adjacent to warm timber glue lams align to create a rhythm showing progressive movement from one end to another. These materials serve as contextual reminders of the campus's vast number of pine trees and expressed Northwest culture. A suspended bridge eases wayfinding from the second-story classrooms in the old building to all new rehearsal spaces. This organization also creates passive student/faculty interactions in naturally occurring ways.

The addition includes two 2,500 square-foot choral and instrumental rehearsal rooms that are placed directly adjacent to one another on the north side of the building to create mass and structural efficiency. Using split faced block, fixed and suspended ceiling reflectors, and movable sound-absorbing panels, the acoustic response of these spaces provide the appropriate acoustical environment that can be adjusted for a variety of groups and instruments.

MAIN STREET/STUDENT SPACE | LARA SWIMMER PHOTOGRAPHY

Twelve smaller teaching/practice rooms, a traditional classroom, a glass-enclosed multi-use flexible space, and administrative offices round out the building's program. There are abundant spaces designed for study, learning, and relaxing equipped with durable furnishings that speak to the mid-century modern flavor.

The overarching challenge of this project was to connect the existing mid-century building to modern times, adhere to the acoustical requirements of the program, and address strict humidity control needs to preserve the instruments and the comfort level of the occupants, all while creating a composition of musical architectural elements that speak to the programs housed within the building. Glass, brick, and scored ground-face block create intentionally sculptural masses and elevations which build spatial rhythms in fenestration, offsets in massing, and changes in plane. Colored and transparent glass curtain walls set in large, vertical splayed openings are arranged to portray a lyrical quality in mullion pattern. Walnut-capped interior patio stairs mimic stage risers for student seating and ensemble performances.

Environmental responsibility is a priority for the University. While LEED certification was not sought, the music building was designed with several sustainable elements that achieved the stated goals. The existing building was repurposed rather than razed, limiting site impact. High content recycled materials were used where possible. Laminate glass was used throughout to minimize solar heat gain while maximizing daylight and acoustical performance.

In addition to the architect and acoustical consultant, the design team included: The Shalleck Collaborative, Inc. (theatre consultant), Kirkegaard Associate (audio systems designer, video systems designer), MW Consulting Engineers (mechanical engineer), Leone & Keeble (general contractor), Integrus Architecture (interior design, structural engineer), Taylor Engineering (civil engineer), and Roen Associates (cost estimating). The reverberation time data were measured in the unoccupied instrumental rehearsal space, and the background noise level data were measured in the unoccupied instrumental rehearsal space with the HVAC system on.

1. Piano Studio 2. Teaching Studio 3. Choral Rehearsal 4. Instrument Rehearsal 5. Percussion Rehearsal 6. Recital Hall 7. Band Room

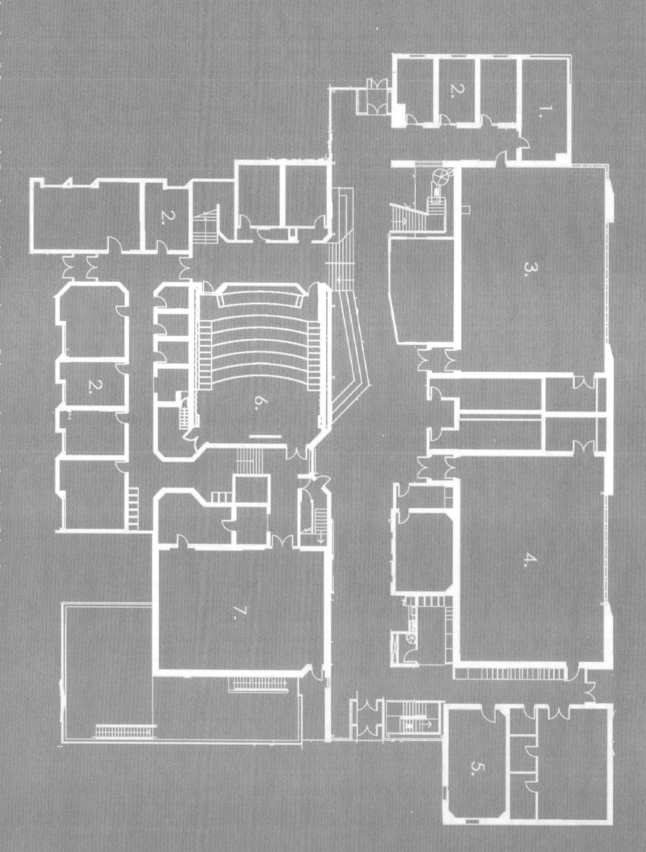

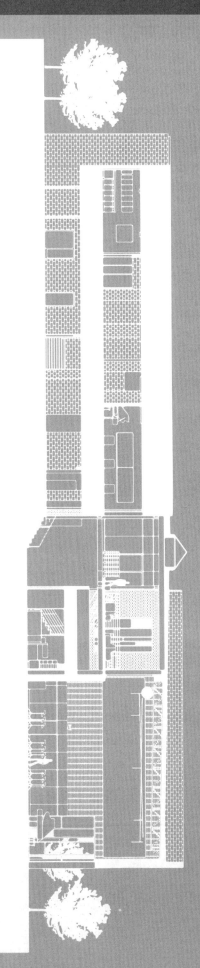

Section

ACOUSTICAL CONSULTANT:
KIRKEGAARD ASSOCIATES

ARCHITECT:
INTEGRUS ARCHITECTURE

COMPLETION DATE:
2017

LOCATION:
SPOKANE, WA | USA

CONSTRUCTION TYPE:
RENOVATION, EXPANSION

CONSTRUCTION/RENOVATION COST:
$9,500,000

FEATURED SPACE DATA:

ROOM VOLUME:
78,336 FT3

FLOOR PLAN AREA:
2,448 FT2

SEATING CAPACITY:
54

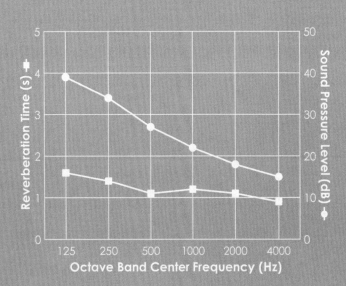

The Adams Center for Musical Arts provides the teaching, practice, and rehearsal facilities that the Yale School of Music needed to carry their program into the future. The new building at the heart of the Adams Center links two historic buildings that the School of Music had occupied for many years - a thorough renovation of the 1895 Hendrie Hall, a handsome Renaissance Revival law school building, and Leigh Hall, a cozy 1930 Gothic Revival health center built for the Law School in 1895 and used for decades by the Yale School of Music. Kirkegaard worked with Butler Rogers Baskett on the 2005 renovation of Leigh Hall (used for teaching studios, offices, and classrooms) and with KPMB on the renovation and addition to Hendrie Hall that created the Adams Center. In addition to the architect and acoustical consultant, the design team included:

Theatre Projects Consultants (theatre consultant), Kirkegaard (audio systems designer, video systems designer), Suzanne Powadiuk Design Inc (lighting designer), AltieriSeborWiebor Consulting Engineers (mechanical engineer), and Consigli (general contractor).

KPMB unearthed the natural beauty of Hendrie by restoring the central staircase and entry corridor, subtly renovating the Yale Glee Club's well-liked rehearsal room, and improving its isolation to the street and to the Yale band rehearsal room above. The Yale band room was transformed by a new skylight and revised acoustic finishes. The rest of the building was completely reconfigured with teaching studios, an opera rehearsal room, and a whole floor of practice rooms. The new classrooms in the building

Glee Club | Kirkegaard

Orchestra Rehearsal Ceiling Reflectors | Kirkegaard

Orchestra Rehearsal Wall | Kirkegaard

are deliberately designed to also serve as ensemble rehearsal rooms.

A new stair tower and elevator was tucked into a light well, making the split-level building fully accessible for the first time. New HVAC systems use chilled beams for quiet, energy-efficient performance.

The five-story addition nearly doubles the size of Hendrie Hall. The heart of the Adams Center is a four-story atrium/lounge that gives the students a central gathering place around its fireplace and easy access to the nearby practice rooms. The percussion and brass faculty have their rehearsal rooms and teaching studios in the new addition, where heavy concrete structure provides excellent isolation from the other spaces.

The jewel of the Adams Center is an elegant new orchestra rehearsal room with warm, clear acoustics and ample natural light. The room includes movable sound-absorbing panels that can be deployed behind wood grilles. Kirkegaard Associates designed a playback/amplification system, built-in video projector and screen, and movable video cameras so the room can easily support distance learning and streaming. Very quiet RC-19 background noise and excellent isolation allow the room to be used for high-quality recordings.

The reverberation time data were calculated for the unoccupied rehearsal room. The background noise level data were measured in the unoccupied rehearsal room with the HVAC system on. The measurements show a faint but noticeable tonal component around 500 Hz from the radiant flooring system. This tonal noise subsequently disappeared, most likely as air in the water lines bled out and the water flow became less turbulent

Opposite page: Orchestra Rehearsal | Kirkegaard

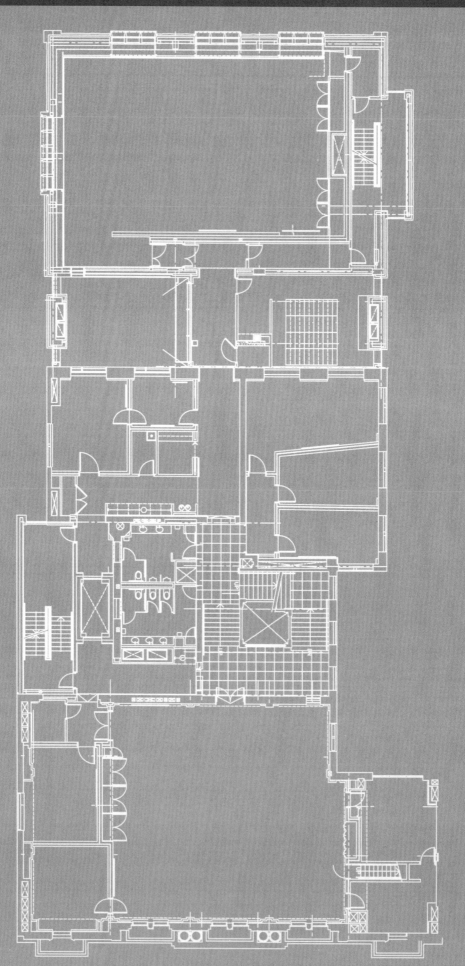

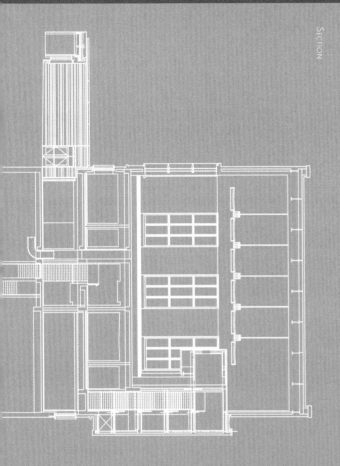

Section

ACOUSTICAL CONSULTANT:
KIRKEGAARD

ARCHITECT:
KPMB ARCHITECTS

COMPLETION DATE:
2017

LOCATION:
NEW HAVEN, CT | USA

CONSTRUCTION TYPE:
NEW CONSTRUCTION, RENOVATION,
EXPANSION

CONSTRUCTION/RENOVATION COST:
$57,100,000

FEATURED SPACE DATA:

ROOM VOLUME:
90,430 FT3

FLOOR PLAN AREA:
2,500 FT2

SEATING CAPACITY:
120-160

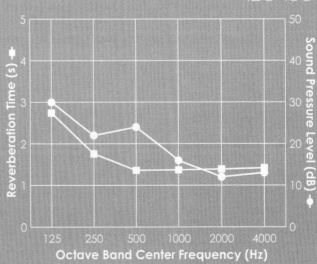

Music Conservatories, Music Rehearsal Spaces, & Community Centers

he Boston Conservatory's urban campus occupies several buildings in the dense Fenway area of Boston. The school had long struggled to provide sufficient space for music practice and rehearsals. When they learned they would lose their long-term lease on off-campus orchestra rehearsal space, they committed to creating a space of their own, in the form of a new building for music and dance rehearsal, practice, instruction, and performance. The site abuts a major highway and rail lines, which leads to potential visibility concerns as well as significant exposure to outdoor noise. The compact site and extensive program resulted in a three-story, 20,000 square-foot building featuring a large orchestra rehearsal room, two dance rehearsal rooms, several large multi-purpose rooms, two teaching studios, and numerous practice rooms, including several ground-floor practice rooms dedicated to percussion. Lobby and upper-floor corridor lounges provide city views.

The largest single space in the building is the orchestra rehearsal room, a 3,230 square-foot trapezoidal room with a height of 24 feet. The ceiling is a grid of acoustical ceiling panels, and pyramidal sound-reflecting panels are interspersed with sound-absorbing panels. Sound absorbing wall panels ring the room at seated and standing head height, and a large curtain provides a degree of acoustical tunability. The room is equipped for informal performances and presentations, and it is used extensively for both, serving as a useful supplement to the school's main music performance space, Seully Hall.

The efficient and compact building layout created a number of sound and impact isolation challenges,

TEACHING STUDIO | CHUCK CHOI

ORCHESTRA REHEARSAL HALL | CHUCK CHOI

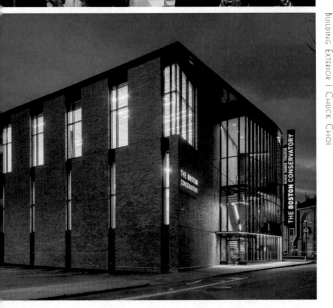

BUILDING EXTERIOR | CHUCK CHOI

most notably between the third-floor dance rooms and the orchestra rehearsal room directly below. A secondary concrete slab was floated on a two-inch isolation system (Kinetics RIM) over the structural composite slab. A spring-supported dance floor system was installed atop the floating slab. The primary concern was impact sound from the dance rooms to the rehearsal room below, and this was adequately controlled: ISR 84 was measured at this location.

Double wall constructions were used extensively, and many rooms, including small practice rooms, included isolated floor assemblies to control flanking sound transmission. In other locations, isolated gypsum board ceilings, framed between structural beams and without contact to the slab above, were employed for vertical sound isolation and control of flanking transmission.

Outdoor site noise from multiple sources – highway vehicles, commuter rail, local street traffic, nearby restaurant exhaust fans – was measured, and resulting interior noise levels were predicted based on planned façade and glazing assemblies. For the most noise-sensitive space, the orchestra rehearsal room, a secondary interior window sash was included to provide improved isolation from outdoor noise.

In addition to the architect and acoustical consultant, the design team included: LAM Partners (lighting designer), WSP Flack + Kurtz (mechanical engineer), Shawmut Design & Construction (general contractor), and LeMessurier (structural engineer). The reverberation time data were measured in the unoccupied space, and the background noise level data were measured in the unoccupied space with the HVAC system on.

Opposite page: Orchestra Rehearsal Hall | Chuck Choi

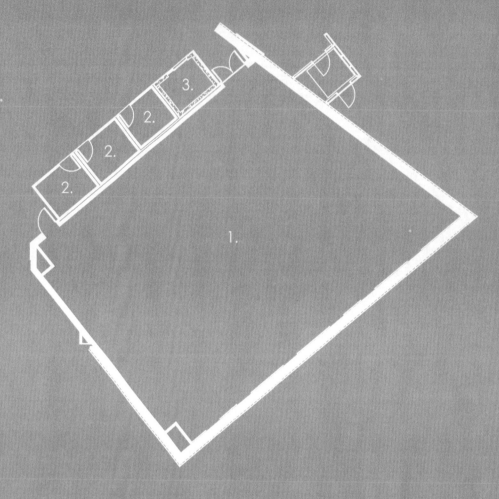

ORCHESTRA REHEARSAL HALL – FLOOR PLAN

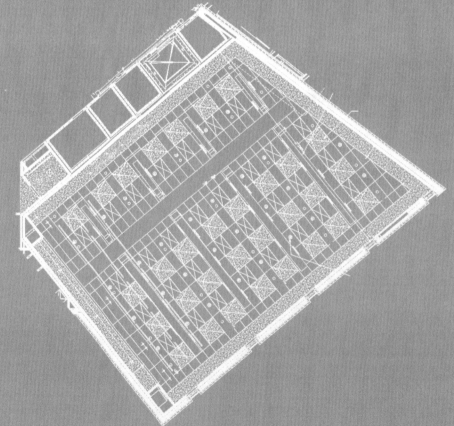

ORCHESTRA REHEARSAL HALL – REFLECTED CEILING PLAN

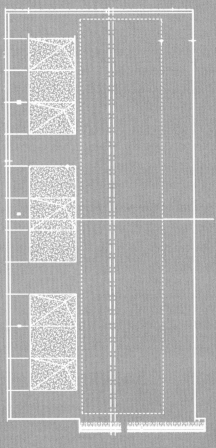

ORCHESTRA REHEARSAL HALL – NORTH ELEVATION

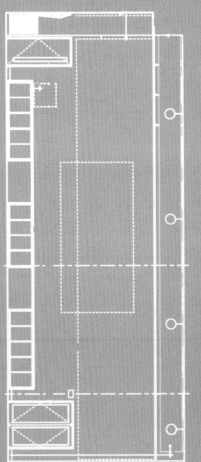

ORCHESTRA REHEARSAL HALL – WEST ELEVATION

ACOUSTICAL CONSULTANT:
ACENTECH INCORPORATED

ARCHITECT:
HANDEL ARCHITECTS / UTILE, INC.

COMPLETION DATE:
2014

LOCATION:
BOSTON, MA I USA

CONSTRUCTION TYPE:
NEW CONSTRUCTION

CONSTRUCTION/RENOVATION COST:
$6,759,850

FEATURED SPACE DATA:

ROOM VOLUME:
77,520 FT3

FLOOR PLAN AREA:
3,230 FT2

SEATING CAPACITY:
130

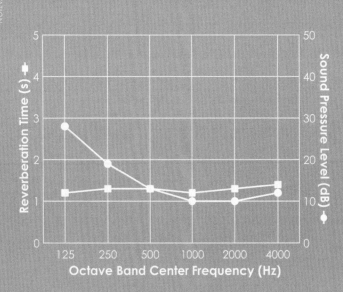

The Curtis Institute of Music is among the highest-ranked music schools in the world. When they needed to expand their facilities to provide additional music education and living spaces for their students, their solution was to replace the Locust Club building with a new building that also incorporated two historical façades. The result is Lenfest Hall, which The Philadelphia Inquirer noted in 2011 "is a remarkable achievement of culture- and, in key places, quiet. [...] Lenfest Hall opens on time, several million dollars under budget, and fully funded." [1]

Lenfest Hall provides Gould Rehearsal Hall for the Curtis Symphony Orchestra; thirty-two additional teaching studios with non-parallel walls, chamber music rehearsal rooms, and practice rooms; audio and video recording studios; an orchestra library; and the orchestral instrument collection, which houses over two hundred string instruments and bows and over fifty wind and brass instruments, all requiring special attention to mechanical systems for ensuring their preservation. The facility provides safe, affordable residences for 80 students – nearly half the student body – and has amenities that include dining and social spaces and a roof terrace, all of which are shared among students, faculty, and staff.

The 3,200 square-foot Gould Rehearsal Hall incorporates lower walls that are lightly angled in section and has horizontally traveling acoustical drapery located at the upper two-thirds of three walls to accommodate rehearsal and performance configurations for ensembles ranging in size from

a soloist with accompaniment to full orchestra. The unoccupied reverberation time of the room ranges from 1.0 second with all absorption deployed to 1.3 seconds with all absorption stored. An array of overhead ceiling reflectors provides supportive reflections for the ensemble while still allowing sound energy access to the upper volume for loudness control. The background noise level data were measured in the unoccupied space with the HVAC operating at 100%.

Since the building had to accommodate simultaneous music-making from many rooms, and students often practice during nighttime hours while others are sleeping, the building's exceptional requirements for sound isolation between rooms and at the building exterior demanded careful attention to nearly every design detail and construction process. For example, the rehearsal hall features a full wall of windows to match the historical façades of the existing Curtis Institute buildings. A double-glazing system with a deep airspace is used for improved isolation from exterior noise. This assembly is also used in all teaching studios and other music-making rooms with exterior windows. The project has achieved LEED Gold Certified status and won numerous awards.

In addition to the architect and acoustical consultant, the design team included: Cosler Theatre Design (theatre consultant), Metropolitan Acoustics, LLC (audio systems designer, video systems designer), Grenald Waldron Associates (lighting designer), Marvin Waxman Consulting Engineers, Inc (mechanical engineer), Keast & Hood Co. (structural engineer), Noble Preservation Services, Inc. (historic preservation architect), and Stantec (civil engineer).

[1] Dobrin, Peter. "Lenfest Hall Opens a New Era for Curtis Institute of Music." *The Philadelphia Inquirer*, 6 September 2011.

Opposite page: Rehearsal Hall Interior - Small Ensemble | Tom Crane Photography

Above: Teaching Studio | Tom Crane Photography

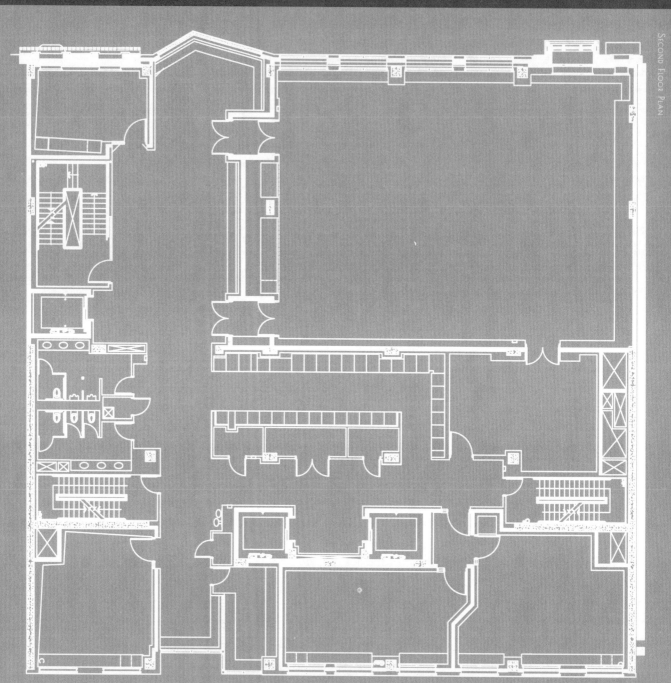

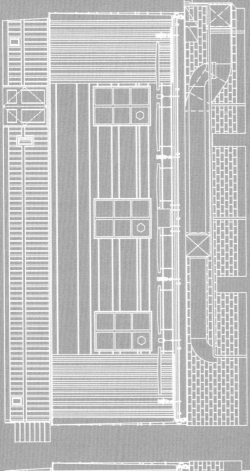

Rehearsal Hall – Longitudinal Section

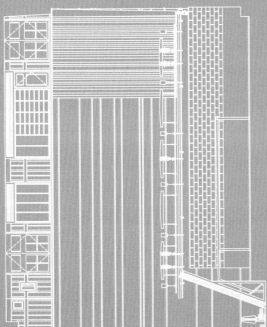

Rehearsal Hall – Transverse Section

ACOUSTICAL CONSULTANT:
KIRKEGAARD

ARCHITECT:
VENTURI, SCOTT BROWN AND
ASSOCIATES, INC.

COMPLETION DATE:
2011

LOCATION:
PHILADELPHIA, PA | USA

CONSTRUCTION TYPE:
NEW CONSTRUCTION

CONSTRUCTION/RENOVATION COST:
$65,000,000

FEATURED SPACE DATA:

ROOM VOLUME:
121,000 FT3

FLOOR PLAN AREA:
3,200 FT2

SEATING CAPACITY:
220

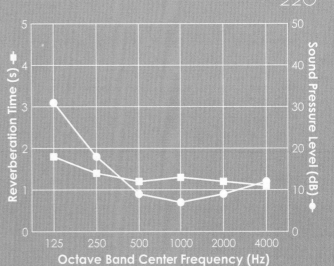

The first meeting with Curtis Institute's organ professor Alan Morrison and pipe organ builder Randall Dyer to inspect former percussion studios in the original Conservatory building (formerly the residence of George & Mary Drexel on Philadelphia's Rittenhouse Square) revealed a warren of four small basement rooms more like dungeon cells than a future organ studio. The acoustic tile ceilings hid a varied assortment of noisy window air conditioners, exhaust fans, and supplemental air transfer systems, plus a hundred-year accumulation of steam, gas and sprinkler pipes, electrical conduit, life safety systems, and more unlabeled phone and data cables than could be counted.

What began with the client's plan for a "paint and carpet" refresh quickly evolved into a major renovation project with areas of selective demolition down to the building structure; however, perimeter walls built for the previous 1980s practice rooms were reused. A particular design challenge was presented by three massive brick piers dividing the studio into three distinct spaces.

The largest room became the teaching studio with organ console (plus a grand piano not mentioned during design that appeared after completion), a smaller corner room was designated as the pipe organ chamber, and a long internal corridor was essential to connect the studio with the main basement hallway. The third original practice room became a sound-isolated mechanical room for new, dedicated HVAC equipment and the organ blower.

From the musician's perspective, the organ studio seems deceptively simple: just the console and a wall of organ pipes. But this "pipe facade" conceals a second room containing the following: the actual pipe organ, multiple windchests for Great, Positiv, Swell and Pedal divisions, air reservoirs, wooden wind trunks, a large wood box around the Swell division with expression (or "swell") shades, service ladders and catwalks, an electronic control system, and of course the sound-producing pipes themselves that range in size from a pencil to rectangular, wood bass pipes one foot wide by more than 8 feet long.

Pipe organs are usually found in large worship and concert spaces having large-room acoustics to match. At Curtis, however, the acoustician was faced with the antithesis, a small room with a low ceiling, and an organ peaking through a small opening in one corner. The client wanted a place where students could practice for hours on end, fully engaged by organ sound, but without suffering listener fatigue so common in small music practice rooms where the only remedy for loud instruments is to maximize sound absorption and minimize the sound of the room itself.

The acoustical solution to this dilemma was found by creating a sound-distribution plenum to connect the organ chamber with the rear of the studio using interstitial space between the upper sound-isolating ceiling and lower, finished studio ceiling, which also hides new and existing building infrastructure from view. This solution required a coupled aperture at the

WALL DIFFUSERS BEHIND ORGAN CONSOLE | DAN CLAYTON

ORGAN CHAMBER (LEFT) AND WALL DIFFUSERS (RIGHT) | DAN CLAYTON

organ chamber and wire mesh sound-distribution ceiling grilles at the studio's side and rear walls.

A secondary acoustical design feature made use of the internal corridor as a sound expansion space by leaving the organ chamber open to the corridor between the aforementioned brick piers. The organ builder proposed locating large wood bass pipes in these openings, so broadband sound diffusers were placed on the opposite wall to minimize sound reflections which might have exaggerated specific pipes and pitches. Studio walls are fitted with broadband diffuser panels alternating with hybrid high-frequency diffuser & mid/low-frequency absorber panels. This combination is designed to scatter sound, balance room frequency response, and prevent boominess and excessive loudness.

While the organ chamber was just large enough for the desired three-manual practice instrument, its location between the mechanical room and studio meant supply-air ductwork had to pass through the organ before heading into the studio. This complex design task was aided by a 3D computer model to make sure all the pieces would fit and function properly. Sound isolation ceilings above the organ chamber and studio prevent sound from traveling up to the first-floor Bok Room, the Curtis Institute's formal meeting and reception space.

This project produced a small organ practice studio which sounds much larger than its actual size, with a pleasant sense of spaciousness and envelopment for organist and listeners. Lateral reflections are partially diffused in the horizontal plane. Typical small-room boominess or mid/low-frequency resonance is well controlled, but without loss of high frequency response. Listeners do not feel aurally compressed by the 8-foot ceiling, but instead feel a sense of height and space overhead from organ sound filtering through the sound-distribution ceiling. It was a lot of work, but all involved in the project agree they started with a pig's ear and turned it into a silk purse!

In addition to the architect and acoustical consultant, the design team included: Marvin Waxman Consulting Engineers (mechanical engineer), INTECH Construction (general contractor), and Randall Dyer & Associates (pipe organ builder). The reverberation time data were measured in the unoccupied space, and the background noise level data were measured in the unoccupied space with the HVAC system on.

Above: Organ Studio | Dan Clayton

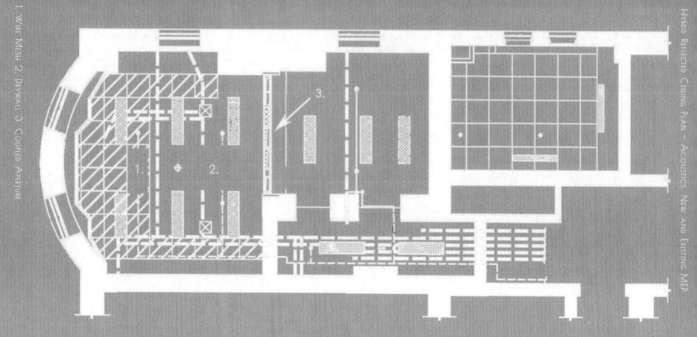

1. Wire Mesh 2. Drywall 3. Coupled Aperture

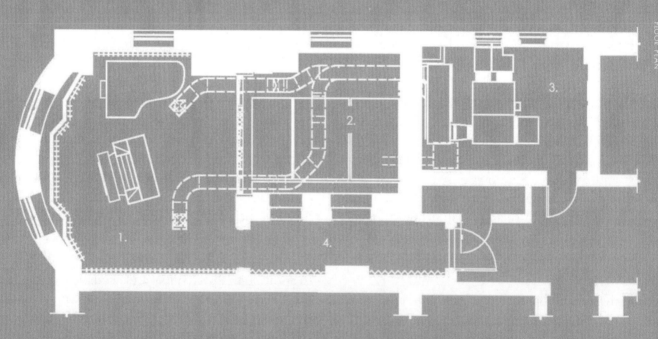

1. Organ Studio 2. Organ Chamber 3. Mechanical Room 4. Studio Corridor

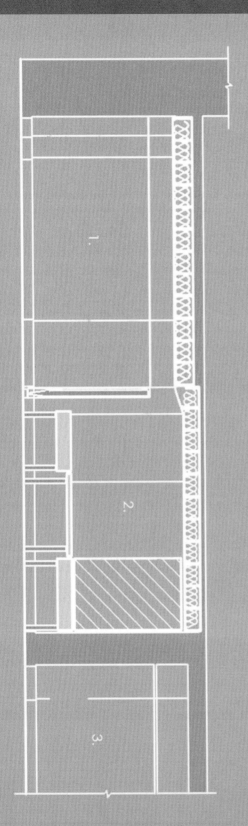

ACOUSTICAL CONSULTANT:
CLAYTON ACOUSTICS GROUP

ARCHITECT:
VSBA ARCHITECTS & PLANNERS

COMPLETION DATE:
2013

LOCATION:
PHILADELPHIA, PA I USA

CONSTRUCTION TYPE:
RENOVATION

CONSTRUCTION/RENOVATION COST:
$650,000

FEATURED SPACE DATA:

ROOM VOLUME:
5,200 FT3

FLOOR PLAN AREA:
530 FT2

SEATING CAPACITY:
15

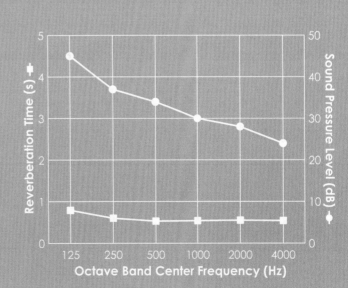

ndian Hill Music's new facility located in Groton, Massachusetts is designed to serve their mission of sharing music through teaching and performance. The non-profit organization consists of a music school, professional orchestra, professional concert series, and community outreach programs. The new building is anchored by a 1,000-seat concert hall that opens to an additional 1,300 seats out on a sloped exterior lawn to yield a performance venue as unique as its rural setting. The facility also delivers 300- and 75-seat recital halls, rehearsal rooms, piano and teaching studios, a percussion studio, a children's music classroom, and a jazz dining room along with back of house support spaces and administrative offices.

The concert hall, fit for a professional orchestra yet intimate enough for young performers and parental audiences, has a graceful, organic form developed through acoustic modeling to iteratively investigate supporting geometries and materials. Wood decking backed with shotcrete form the walls and ceilings to provide full-spectrum overhead reflections, and a canopy of acrylic reflectors above the stage provides in-time reflections for intimacy and excellent ensemble conditions. The canopy is variable in height to accommodate smaller ensembles as well as large orchestral performances. Glulam "wishbones" as they are endearingly called by their shape, form the structural frame and give shape to the sidewalls of the room which are a composite make-up of glazing, stone, and the concrete-backed T&G wood planks.

Clerestories at the upper volume and transparent walls at the ground floor level bring in natural light and provide a connection to the agrarian environment outside. Blackout shades can be deployed when lighting needs to be controlled.

The lower upstage wall is surrounded by diffusion made up of 3 rows of ¼" thick hardwood in a complex pattern of straight, twisted, and curved strips inspired by meadow grasses in a light breeze. On the side walls, where much less diffusion is required, the same pattern is milled from a single layer of ¾" material and glued directly to the concrete. Retractable orchestra risers that store within the stage lifts when not in use further enhance intimacy and ensemble on the stage. The reverberation time in the concert hall is adjustable by acoustic banners made of wool serge at the sidewalls and an upstage concealed velour drape deployed behind the upstage diffusion during amplified performances. The reverberation time data shown were calculated for the unoccupied space. Acoustic modeling and subjective listening to auralizations helped to facilitate the ceiling height, canopy size, side wall shaping, and variable acoustic systems.

Behind the acoustically transparent projection surface is a digital organ that uses sampling technologies to reproduce the sounds of historic period instruments. The organ console is moveable with connection points on the stage to accommodate recital, concerto, and orchestral locations on the stage.

Audio and video systems inside the concert hall

Concert Hall Rendering | Epstein Joslin Architects

Recital Hall Rendering | Epstein Joslin Architects

include four 30,000 lumen projectors, recessed in sidewalls and operated in unison, to map an immersive video environment onto the upstage projection surface. The projectors allow for digital mapping and for projecting video art, images, and text, including supertitles. When the operable rear wall is opened onto the lawn, two exterior line arrays amplify performances on the stage for the lawn audience. Infrastructure is in place for two large video screens and surround sound for an immersive listening experience under the open sky.

Displacement ventilation with "swirl" diffusers delivers slow, quiet airflow from a plenum below the seats, served from mechanical equipment located in the basement below grade. The MEP systems are designed to be undetected in the RC-15 space. The noise rating was calculated for the unoccupied space.

The recital hall has a similar architecture to the concert hall, albeit reduced in scale. With 300 seats it will allow young performers the opportunity to perform in a smaller and more intimate space. The ceiling over the stage is diffusive and lower in height than the rest of the T&G plank ceiling, allowing the omission of a separate canopy. Glass upstage and side walls are subdivided with deep mullions for a

measure of diffusion and follows the "wishbone" shaping present at the sidewalls. A glulam "wishbone" structure with walls and ceiling of concrete-backed T&G plank provides the same massive and reflective construction as the recital hall's big brother. Many of the ensemble and practice rooms also take advantage of exposed wood plank as an acoustic finish, with treatments of felt and tectum to address focuses of the undulating ceilings.

In addition to the architect and acoustical consultant, the design team included: Theatre Consultants Collaborative (theatre consultant), Threshold Acoustics LLC (audio systems designer, video systems designer), Ripman Lighting Consultants, Inc. (lighting designer), Bard Rao + Athanas Consulting Engineers, LLC (mechanical engineer), and Goguen Incorporated (general contractor).

Above: Indian Hill Music Center Exterior Rendering | Epstein Joslin Architects

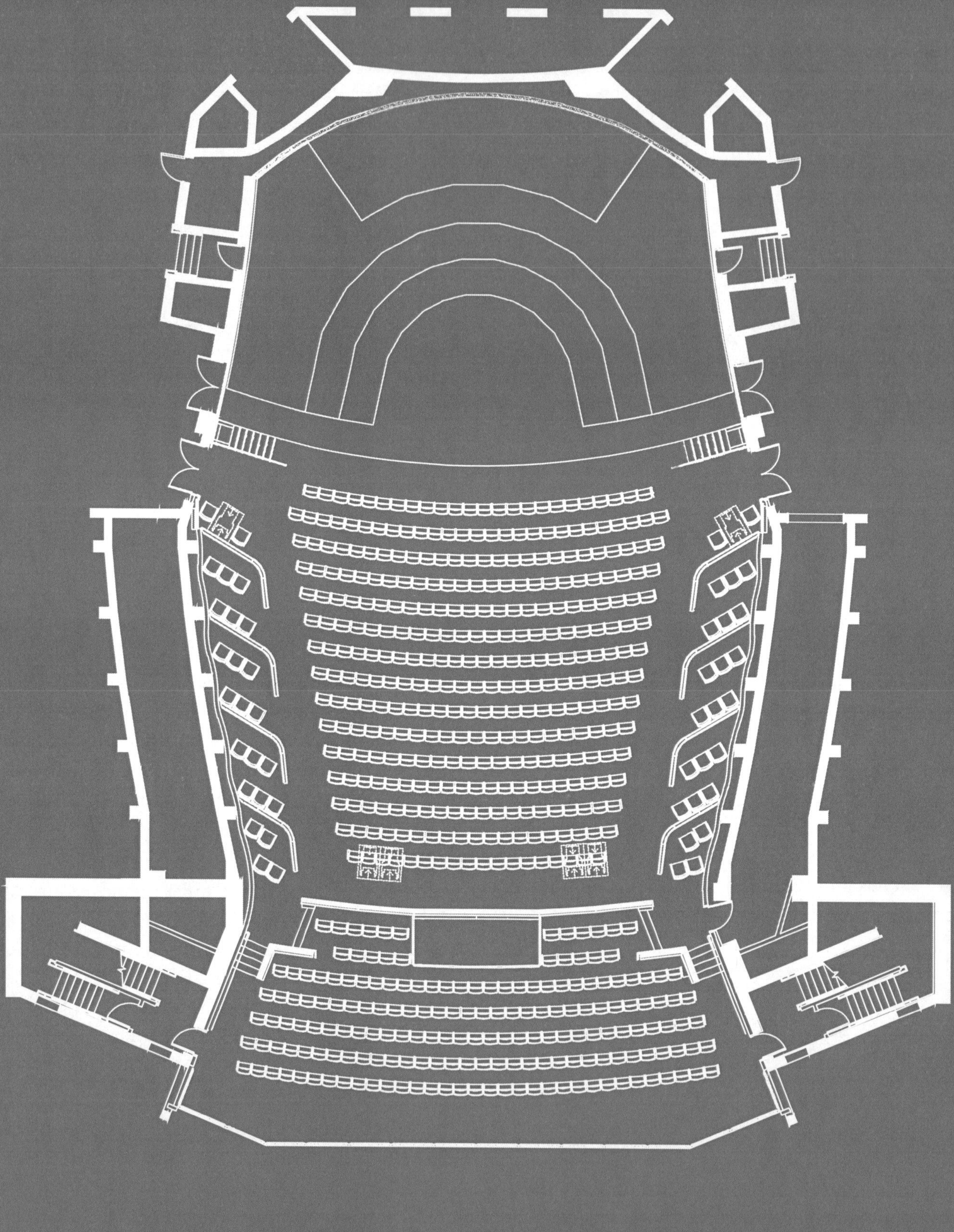

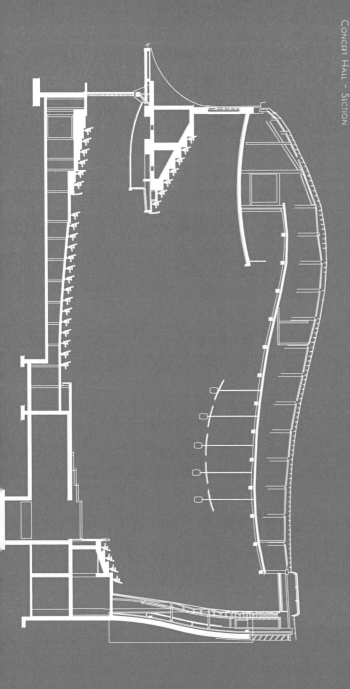

CONCERT HALL - SECTION

ACOUSTICAL CONSULTANT:
THRESHOLD ACOUSTICS LLC /
LKACOUSTICS DESIGN

ARCHITECT:
EPSTEIN JOSLIN ARCHITECTS

COMPLETION DATE:
2020

LOCATION:
LITTLETON, MA I USA

CONSTRUCTION TYPE:
NEW CONSTRUCTION

CONSTRUCTION/RENOVATION COST:
NOT AVAILABLE

FEATURED SPACE DATA:

ROOM VOLUME:
500,000 FT3

FLOOR PLAN AREA:
7,800 FT2

SEATING CAPACITY:
1,037

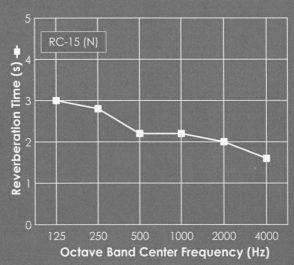

MANHATTAN SCHOOL OF MUSIC

The Manhattan School of Music is a private music conservatory on the Upper West Side of New York City. The school offers degrees on the bachelors, masters, and doctoral levels in the areas of classical and jazz performance and composition. The conservatory has trained some of the world's most celebrated performing artists and has upheld a tradition of excellence in music education for nearly a century.

Neidorff-Karpati Hall, formerly Borden Auditorium, reopened in November of 2018 after undergoing a complete interior renovation, including updated decor with new lighting, seats and carpeting; a widened proscenium; a shortened balcony; a brand new wooden acoustic shell and a forestage reflecting canopy. Jaffe Holden provided acoustic consultation

and audio/video systems design services for this project. The design team also included: Schuler Shook (theatre consultant, lighting designer), Kohler Ronan Consulting Engineers (mechanical engineer), and Yorke (general contractor).

A completely new technical infrastructure was created with a state-of-the-art sound and recording system. Finishes were hardened to enhance sound reflection, and diffusive elements were added to the side walls to better balance sound. An expanded musical theater program required extensive use of the sound system, so an adjustable acoustic banner system was installed on the side walls to lower reverberation time and increase speech intelligibility.

The existing stage was fraught with acoustical challenges. It was small and could not accommodate

the growing number of enrolled musicians. Space was extremely limited, so there was no space for a typical shell. Nearly a quarter of the students were hidden in the wings between the curtains in an extreme thrust stage condition that pushed musicians into the house without the benefit of any acoustic reflection support. Thus, the sound on stage was too loud and lacked resonance and onstage hearing.

To overcome these limitations, the acoustical consultant opened up the proscenium by enlarging its width. Small support columns remained in place and contributed to a unique design of proscenium columns within the stage area. A new shell was designed in three pieces. The rear wall piece is a single unit that flies up to allow circulation underneath and upstage. The two side wall pieces travel across the stage on motorized tracks. For tuning purposes, the lower portion of these walls can rotate open in various degrees to vent overly loud sound from brass and percussion. The ceiling panels are the traditional tip and fly units but also have a series of gaps in them to properly balance sound and reduce loudness. The new forestage ceiling array provides sound reflections for musicians located out on the stage extension and helps project sound to the audience.

Improvements in the house included a new, triple layer gypsum board ceiling, new sound and light locks, and removal of three balcony rows to improve sightlines and under-balcony acoustics. These improvements increased the reverberation time from the old and rather dry 1.5 seconds to 2.0 seconds unoccupied. The adjustable banner system and rear wall walk-along drapes can lower the reverberation time back down to 1.5 seconds. The background noise level data were measured in the unoccupied space with the HVAC system on, and the noise ratings were calculated for the unoccupied space.

When the hall is used for opera or musical theater, a portal set is lowered from the fly loft and erected to create a proscenium. Combined with a new sound and recording system and infrastructure to reinforce musical theater performances, jazz bands, and pop artists, the new hall matches the quality of the immensely talented student artists who come to study at the conservatory.

Opposite page: Neidorff-Karpatti Hall - Stage View | Jaffe Holden Acoustics

Above: Musicians Rehearsing on Stage | Jaffe Holden Acoustics

MANHATTAN SCHOOL OF MUSIC

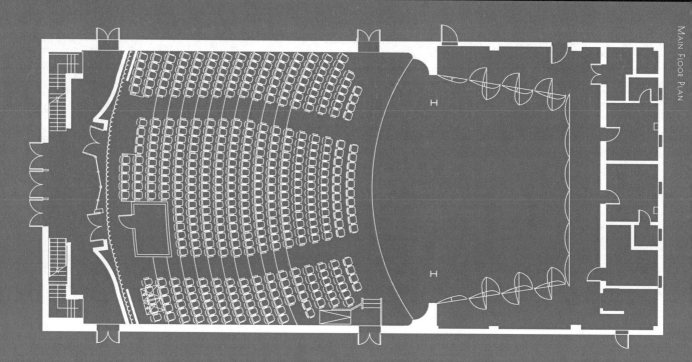

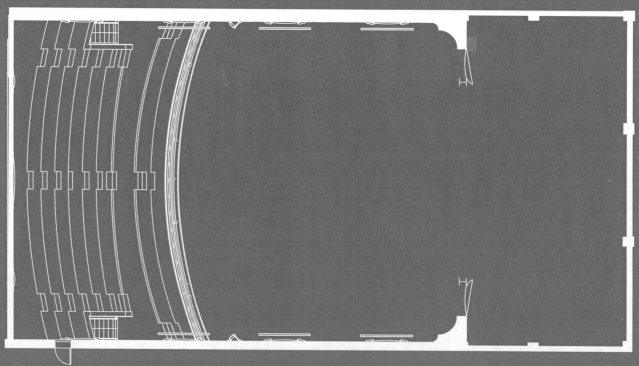

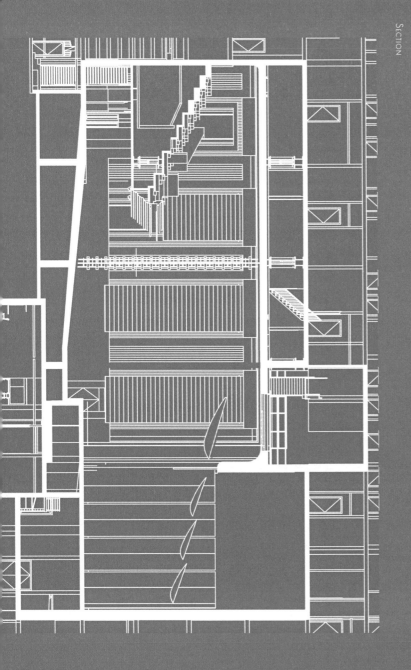

SECTION

ACOUSTICAL CONSULTANT:
JAFFE HOLDEN ACOUSTICS

ARCHITECT:
HOLZMAN MOSS BOTTINO ARCHITECTS

COMPLETION DATE:
2018

LOCATION:
NEW YORK, NY | USA

CONSTRUCTION TYPE:
RENOVATION

CONSTRUCTION/RENOVATION COST:
$6,034,440

FEATURED SPACE DATA:

ROOM VOLUME:
175,000 FT3

FLOOR PLAN AREA:
7,500 FT2

SEATING CAPACITY:
650

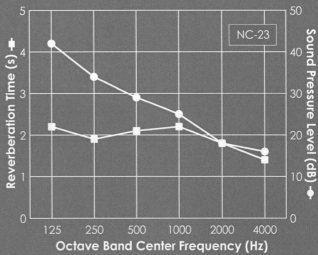

The abilities of young musicians attending Mount Royal Conservatory of Music far exceeded the quality of the former recital space on campus. In 2005, TALASKE was hired to perform a complete survey of the music education facilities. This set the stage for the 2009 design of the new 787-seat Bella Concert Hall, a recital/performance space, dozens of teaching studios, and the Music with Your Baby program. TALASKE continued as the design acoustician for the project. In addition to the architect and acoustical consultant, the design team included Auerbach Pollock Friedlander (theatre consultant).

The need for acoustic volume was well accommodated by the architect's desire to create a tall and slightly-peaked room to reflect the Canadian Rocky Mountains in the background. The high ceilings necessitated lower sound reflecting panels and the province's state flower, the Alberta Rose, provided the inspiration for the panels above and forward of the stage. The design is complemented by side balconies and intermediate shelves, shaped for achieving favorable sound reflections offering *clarity* and *spaciousness*.

The Bella Concert Hall serves as the primary performance space of the Conservatory and is the second home for The Calgary Philharmonic Orchestra.

Unlike some designers who determine the acoustic volume requirements based on the number of patrons, the acoustic volume of the hall was set based on the number of musicians expected on stage. The stage size and stage proportions are set to accommodate the range of expected music ensembles including up

to 85-musicians on the platform and 125 choristers in the choral loft. The acoustic volume is approximately 500,000 cubic feet.

With one exception, in acoustic music mode the only sound absorbing finishes within The Bella relate to audience seating, musicians, limited carpet areas, and patrons. Non-porous surfaces, in combination with an optimized shape and dimensions, result in a sound reflection pattern offering clarity of sound, envelopment, reverberance, and running liveliness. Heavy concrete materials were selected to minimize absorption of low-pitched sound. Similarly, surfaces were multi-layer in design and tied back to the concrete wall, also to avoid low-pitched sound absorption. When low-pitched sound is retained, the lower register instruments resound and offer a warmth of sound which is highly desirable.

The exception to abundance of sound reflecting materials is a custom retractable absorption system which is integrated into the architectural design of the room. Controlled by a dedicated computer system, sound absorbing curtains, banners, and panels move out from pocketed enclosures. The amount of exposure for sound absorbing finishes depends on the intended performance type and the retractable absorption system was optimized during the commissioning process. Upon pressing a single button, the room acoustic environment changes in a prescribed manner to optimize sound conditions for each different ensemble on stage. Settings for rehearsals were also programmed.

The design of the room was optimized using an acoustic computer model. The room was digitally defined in 3-D space and acoustic attributes for each surface were identified. The shape of walls, balconies, shelves, ceilings, seating, and all other surfaces were identified. Sound reflective, sound absorptive, or sound diffusive surfaces were identified accordingly, and these acoustic properties were accounted for during the subsequent processing of the computer model.

Once the model was assembled, the timing, direction, and strength of thousands of sound reflections were simulated. The resulting impulse response was convolved with music recorded within an anechoic space. Then-dean Dr. Paul Dorian listened to the results and contributed to the process.

The tall ceiling dictated the need for suspended sound reflecting panels. These were quite large and heavy to reflect low-pitched sound in addition to mid- and high-pitched sound. While optimized within the computer design, the precise angle of these panels needed to be set properly during construction to map fully sound from the entire stage to select seating sections. Lasers were used to precisely set/confirm the angles of these surfaces during the construction process.

The mid-frequency Early Decay Time within the unoccupied hall was measured to range from 1.7 to 3.1 seconds for all retractable absorption exposed and retracted, respectively. The mid-frequency T30 values within the unoccupied hall were measured and range from 1.3 to 2.6 seconds for all retractable absorption exposed and retracted, respectively. C80 values within the unoccupied hall were measured to range from +1 to -3dB for all retractable absorption exposed and retracted, respectively. The resulting background noise level was approximately RC-17.

Opposite page: View from House Left Balcony | R. Talaske

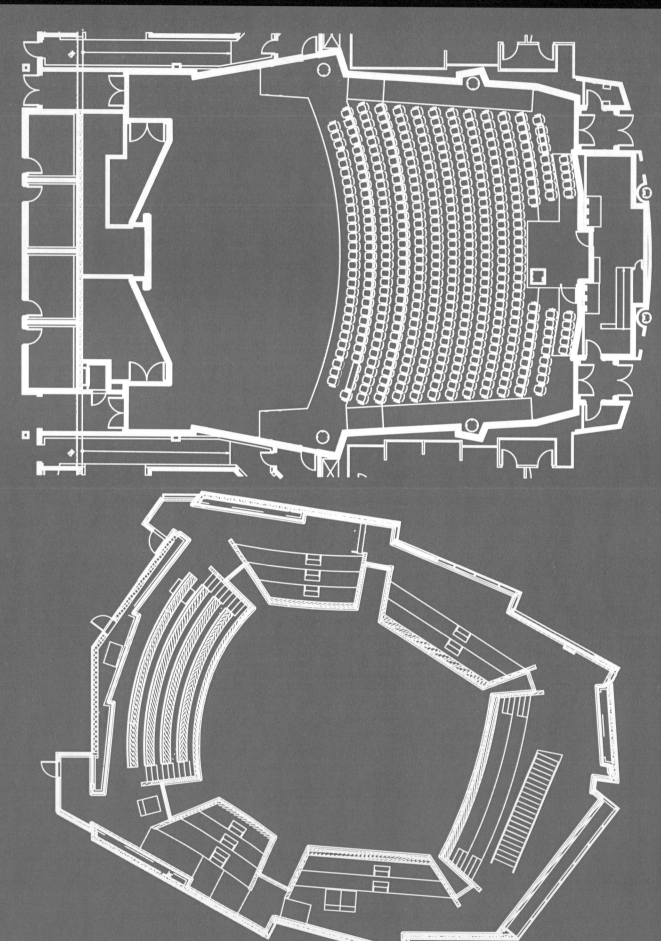

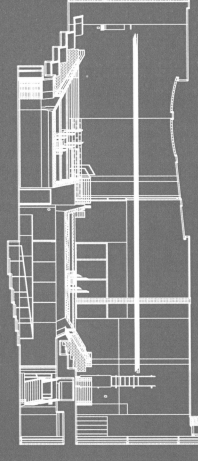

LONGITUDINAL SECTION

CROSS SECTION

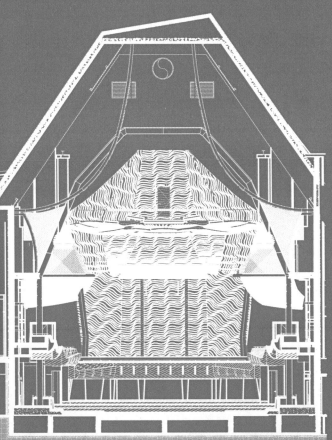

ACOUSTICAL CONSULTANT:
TALASKE | sound thinking

ARCHITECT:
PFEIFFER PARTNERS

COMPLETION DATE:
2015

LOCATION:
CALGARY, ALBERTA | CANADA

CONSTRUCTION TYPE:
NEW CONSTRUCTION

CONSTRUCTION/RENOVATION COST:
$60,000,000

FEATURED SPACE DATA:

ROOM VOLUME:
500,000 FT3

FLOOR PLAN AREA:
7,414 FT2

SEATING CAPACITY:
787

The Music Academy of the West is a preeminent school for gifted young classically-trained musicians from around the world. At its ocean-side, ten-acre campus in Santa Barbara, California, the Academy provides these musicians with the opportunity for advanced study and performance under the guidance of internationally renowned faculty artists, guest conductors, and soloists.

The new Hind Hall Teaching Studio Building is located in the center of the campus. It houses two large ensemble rehearsal rooms - the Gondos String Quartet Studio (19ft x 21ft) and the Lehrer Percussion Studio (20ft x 33ft) - plus six large teaching studios. Studio ceiling heights were maximized within campus building constraints to 13ft. The building features a new central courtyard, convenient loading dock, and controlled environments for instrument storage.

All music rooms are isolated with full box-in-box construction. All windows and doors are custom wood fabrications, designed for suitably high sound isolation. Doors feature cam-lift hinges and full vision panels. In each room there is at least one double door large enough to move a grand piano. Ductwork was designed to minimize crosstalk and intrusion from outdoor noise.

The HVAC system provides carefully-controlled temperature and humidity, especially since all rooms except the Percussion Studio house Steinway grand pianos. The HVAC in all rooms was measured to operate at quieter than NC 20.

Each room is individually treated for optimal sound

GONDOS STRING QUARTET STUDIO | MEHOSH PHOTOGRAPHY

LEHRER PERCUSSION STUDIO | MEHOSH PHOTOGRAPHY

CUSTOM SOUND ISOLATION WINDOW

SOUTH ENTRANCE

TYPICAL INTERIOR

diffusion, balanced with scattered but relatively minimal sound absorptive treatment. Treatments combine absorptive and diffusive panels having the same fabric finishes. Windows and doors provide some modest diffusion, and their locations were considered in positioning the wall treatments. Two adjacent walls in each room are slightly angled, minimizing flutter echoes and resultant coloration. The results are rich acoustics with extreme clarity.

All rooms have moderately longer low-frequency reverberation compared to the middle frequencies for added warmth, but also have slightly longer high-frequency reverberation for some sparkle and liveliness. The measured reverberation times are fairly consistent, except that the Gondos String Quartet Studio was designed to have slightly longer reverberation times.

All rooms feature an array of absorber/reflectors in the ceiling. The ceilings of the two larger studios add 4ft x 4ft doubly-curved diffuser/reflectors that are hidden behind a scrim of sound transparent fabric to maintain a visually level ceiling, but without adverse effects.

The Percussion Studio needed a large storage closet for the variety of permanently housed instrumentation, which presented an opportunity. The wood folding doors to this closet cover much of one end wall, and are designed to be highly transmissive, thus allowing some sound energy to be conveniently discarded into this added volume, thereby helping to tame the overall sound levels.

With the consistent sound quality of each teaching studio and the purposely-adjusted responses of the two larger rooms, Hind Hall is in constant high demand by both educators and students.

SOUND-TRANSMISSIVE PERCUSSION CLOSET DOOR

1. Studio 2. Percussion Studio with Storage Closet 3. String Quartet and Ensemble

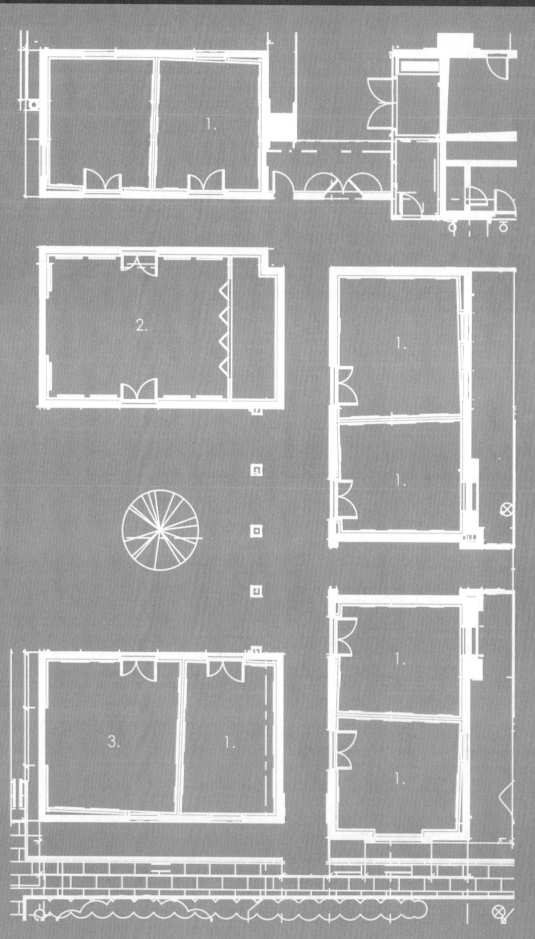

ACOUSTICAL CONSULTANT:
MCKAY CONANT HOOVER INC.

ARCHITECT:
PMSM ARCHITECTS
(NOW 19SIX ARCHITECTS)

COMPLETION DATE:
2017

LOCATION:
SANTA BARBARA, CA | USA

CONSTRUCTION TYPE:
NEW CONSTRUCTION

CONSTRUCTION/RENOVATION COST:
$4,400,000

FEATURED SPACE DATA:

ROOM VOLUME:
3,180 FT3 TO 8,450 FT3

FLOOR PLAN AREA:
265 FT2 TO 660 FT2

SEATING CAPACITY:
AS NEEDED

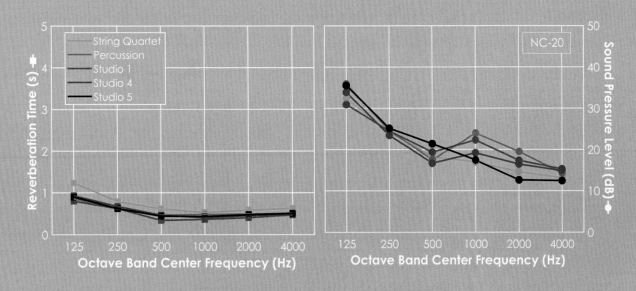

The Musikinsel, or Island of Music, in Rheinau Switzerland is a former Benedictine monastery located on the Rhine River. Its complex, a protected building, was converted into a music school and a rehearsal venue for orchestras, academies, and chamber music master classes among other events. With an overall context prerequisite of "Silence," the main challenge was to establish optimal acoustics without disrupting the charismatic architecture of the location. Another design challenge was to highlight the inspiring atmosphere in order to offer the musicians full concentration in any of the 16 rehearsal rooms. The rehearsal rooms include small individual practice rooms, medium-sized rooms for 15 people, two large rehearsal rooms, and the music room for performances. The school also maintained an old library. The large rehearsal rooms and performance room cover approximately 200 square meters each and can hold a full orchestra and a big choir.

Kahle Acoustics played an active part in the creative collaboration of the design team led by Bembé Dellinger architects and also included Mr. Beat Zoderer (artist), Mrs. Annette Douglas (textile designer), and Reflexion AG Switzerland (lighting designer). Altogether, the acoustical consultants embraced the dual challenge of preserving the architecture and ensuring the functionality of each practice room irrespective of whether they would be used separately or simultaneously.

The optimized acoustics scope included both building acoustics and room acoustics. The acoustical consultants were very fortunate to work

REHEARSAL ROOM – CARPET, CURTAIN & WALL PANELS | KFB POLSKA

HISTORICAL REHEARSAL ROOM WITH CARPET | KFB POLSKA

WALL PANELS: HIDDEN CLADDING IN RADIATOR | KFB POLSKA

ACOUSTICALLY ACTIVE FURNITURE | KFB POLSKA

Mobile Panels | KFB Polska

Musikinsel Rheinau on the River Rhine | KFB Polska

Musiksaal (Music Hall) | KFB Polska

with half-meter to one-meter thick monastery walls. This existing wall design already eliminates the transmission of loud music between individual practice rooms. However, within rooms, the landmarked architecture did not leave much leeway for an acoustic intervention. The acoustical consultants developed custom fabrics along with acoustically active furniture in order to preserve the architecture. For instance, acoustic absorption was applied in the rehearsal rooms with mobile panels, carpets on parquet floors, curtains, and integrated absorption in the radiator covers. The combination enhances control of the reverberation time and leaves musicians free to adapt the acoustics to their preference.

There is a historical ceiling that provides diffusion, and the reverberation can be tamed with fixed panels behind a piano and a floor carpet on the wooden floor. The vaulted ceiling is acoustically interesting as it produces mainly positive focusing effects due to its concave curvature. The mobile panels are made of sound-absorbing cushions on one side yet are sound-reflecting on the other side, providing flexibility that can support orchestras amongst various other ensembles.

Our largest acoustic intervention was in the Musiksaal, or main music room, as it did not contain any historic fabric. The acoustical consultants constructed the highest possible ceiling height and created a coffered ceiling slightly tilted to prevent flutter echoes and to offer good sound distribution through reflected sound. In addition, the acoustical consultants covered one full wall with acoustic panels that alternated between absorbing and reflective to offer partial reflection back to the orchestra. Lightweight and semi-transparent micro-porous curtains provided wall absorption and were used to create a dryer acoustic environment, especially in the case of large orchestra rehearsals. The reverberation time data were measured in the unoccupied space, and the noise ratings were calculated for the unoccupied space.

The Musikinsel Rheinau reopened in 2014 and has been a very dynamic and inspiring spot for practicing musicians and music education.

MUSIKINSEL RHEINAU

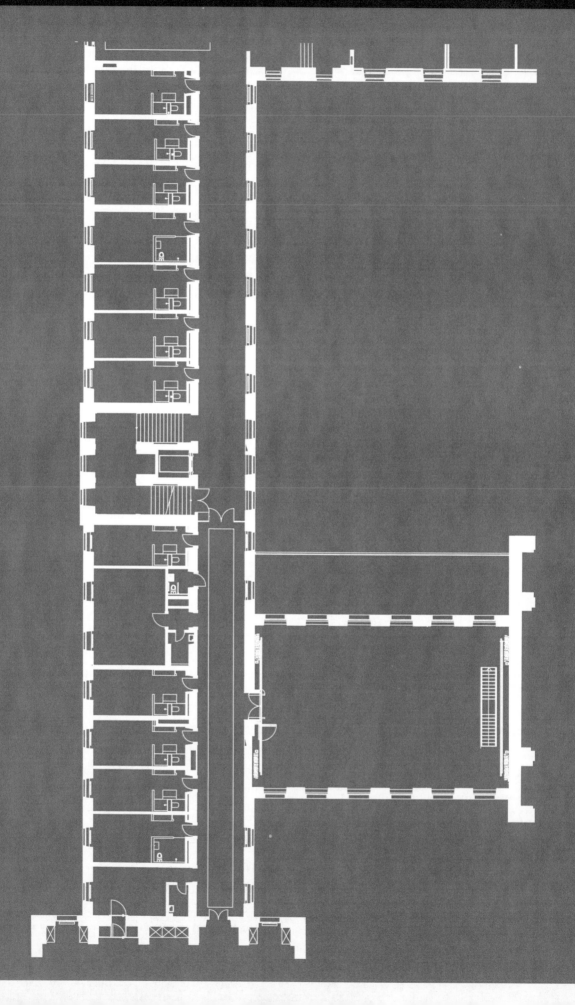

ACOUSTICAL CONSULTANT:
KAHLE ACOUSTICS

ARCHITECT:
BEMBÉ DELLINGER ARCHITEKTEN BDA UND
STADTPLANER

COMPLETION DATE:
2014

LOCATION:
RHEINAU I SWITZERLAND

CONSTRUCTION TYPE:
ADAPTIVE REUSE

CONSTRUCTION/RENOVATION COST:
$28,240,800

FEATURED SPACE DATA:

ROOM VOLUME:
69,217 FT3

FLOOR PLAN AREA:
2,637 FT2

SEATING CAPACITY:
120

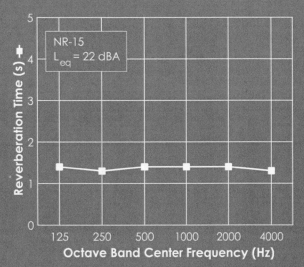

Fifty Oak Street the home for the San Francisco Conservatory, is inserted into the shell of a historically significant Catholic Boys Club in the Civic Center performing arts district. A substantial eight-story addition, with two of those stories underground, supplements the original building. The comprehensive programming includes teaching studios, practice rooms, classrooms, offices, and three major spaces that accommodate both rehearsals and performances.

The 160-seat Sol Joseph Recital Hall offers an intimate, clear, supportive setting for traditional recitals. The hall has a narrow balcony that wraps around three sides with one row of seating. There are adjustable acoustic banners to control reflections and reverberation time with exposed technical catwalks

in the upper volume. The Osher Salon is perhaps the best-loved space in the building – an elegant flat-floor rehearsal room that can seat up to 120 and welcomes less-formal recitals. The upper floors within the original building are organized around an atrium that is sprinkled with social space and study space for the students – an important amenity in an urban school. Nooks and niches at every level of the atrium are popular with the Conservatory's many guitar students.

The ornate ballroom of the Boys Club was repurposed as the audience chamber of the 450-seat concert hall. The original historic interior was maintained, and acoustical treatments were carefully integrated into the historic fabric. A new stage house, equipped with an orchestra pit lift and adjustable acoustic banners,

CONCERT HALL - INTERIOR FROM AUDIENCE

CONCERT HALL - AUDIENCE CHAMBER SIDEWALL

CONCERT HALL - STUDENT REHEARSAL

was built in the addition. The space had to be heavily isolated from the dense surrounding programming by utilizing multiple floating assemblies, massive construction, and isolation joints throughout the building.

The original coffered ceilings with ornate molding along with pilasters and cornices break up the walls and ceiling into complex shapes. These features provide useful medial and lateral reflections to the audience chamber while controlling flutter and echoes. A grid above the stage integrates concealed lighting while providing well-timed reflections for onstage communication for students and performers. A large convex-shaped lid for the stage house sits above the lighting grid and is partnered with the gentle shaping at the stage walls. Together they provide projective geometry to improve on-stage communication for musicians. There is additional volume above the visual ceilings of the stage and audience chamber for loudness control.

The multiple banners in the audience chamber and onstage can be exposed or pocketed to provide a wide range of reverberation time. This range in decay time, provided by the variable acoustics, supports the many uses of this space for classes, recitals, and lightly staged opera productions as well. The thoughtful blend of old and new has made the concert hall successful as the primary rehearsal and performance space for the conservatory's large ensembles as well as a popular rental space for events and touring performances.

The reverberation time data were measured in the unoccupied concert hall. The background noise level data were measured in the unoccupied concert hall with the HVAC system on, and the corresponding noise ratings were calculated for the unoccupied space. In addition to the architect and acoustical consultant, the design team included Auerbach Pollack Friedlander (theatre consultants, audio systems designer).

Opposite page: Concert Hall - Interior From Stage

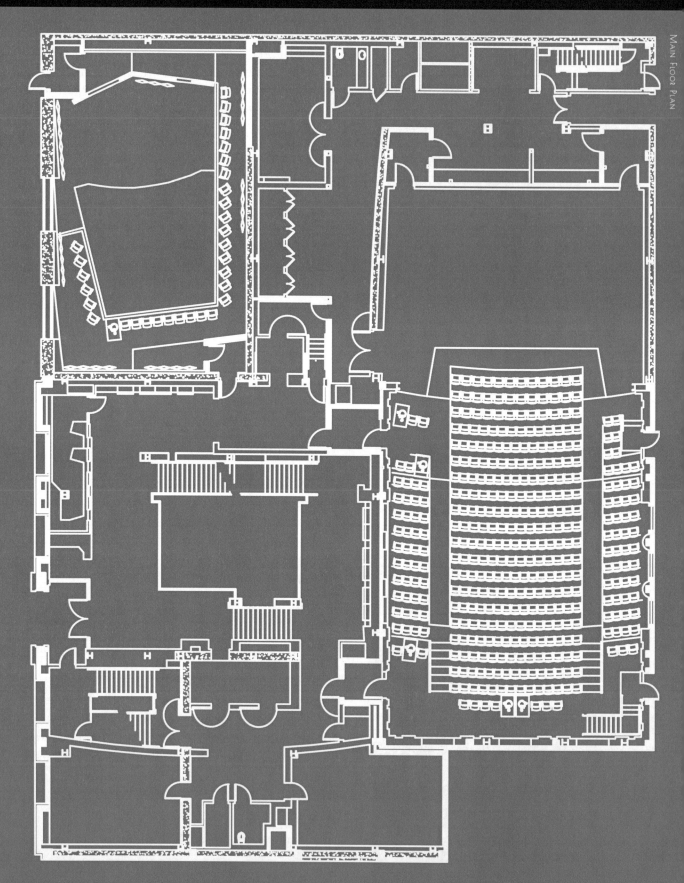

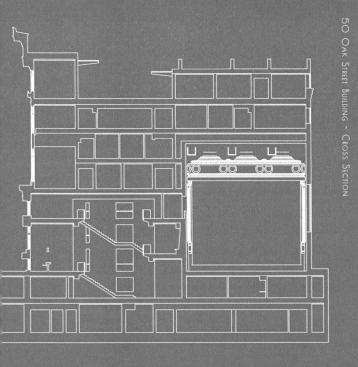

50 Oak Street Building – Cross Section

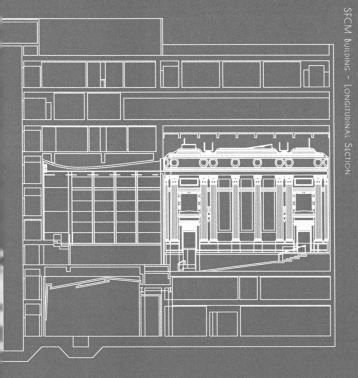

SFCM Building – Longitudinal Section

ACOUSTICAL CONSULTANT:
KIRKEGAARD

ARCHITECT:
PERKINS + WILL

COMPLETION DATE:
2006

LOCATION:
SAN FRANCISCO, CA I USA

CONSTRUCTION TYPE:
NEW CONSTRUCTION, ADAPTIVE REUSE

CONSTRUCTION/RENOVATION COST:
$43,000,000

FEATURED SPACE DATA:

ROOM VOLUME:
218,250 FT3

FLOOR PLAN AREA:
5,160 FT2

SEATING CAPACITY:
450

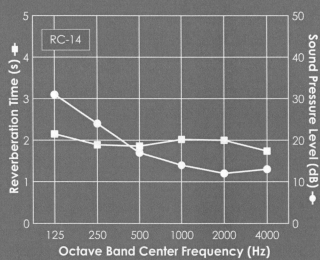

From 1992 to 1999, Kirkegaard contributed to a comprehensive reworking of the facilities for the University of Cincinnati's College-Conservatory of Music (CCM). In addition to the architects and acoustical consultant, the design team included Theatre Projects Consultants (theatre consultant). The first step was the conversion of a 100-year-old gymnasium into the Dieterle Vocal Arts Center that consisted of a choral rehearsal room, choral teaching studios, and a large opera rehearsal room. The second step was the conversion of a 1920s dormitory into teaching studios and practice rooms. The adjacent 1960s music and theater building was then lightly renovated and enhanced with the addition of dance studios and a new scene shop.

The next step in creating a comprehensive suite of music facilities for CCM was to renovate the existing Corbett Auditorium. The venue serving as CCM's primary performance venue was a drab, scale-less 750-seat proscenium theater clad in Formica and suffering from a clattery, overly bright sound with excessive background noise.

The final step was the replacement of Mary Emery Hall with a new three-story building over a parking garage. In addition to a variety of classrooms, rehearsal rooms, offices, and teaching studios, the building houses the beautiful 250-seat Werner Recital Hall, a 100-seat master classroom, and the elegant, flexible, 120-seat Cohen Family Studio Theater, which is used for drama, musical theater, and opera.

Corbett Auditorium was gutted and its interior completely rebuilt. The 3/4-inch thick Formica panels

Corbett Auditorium Post-Renovation

Cohen Family Theater | Kirkegaard

Werner Recital Hall

that covered the walls were removed to expose the rough sprayed-on concrete, which was then sealed with paint. At the rear wall the folded plate diffusion was replaced with Schroeder diffusers and bowed surfaces. The resulting highly diffusive rear wall is excellent for clarity in the audience area but is too heavy-handed for room response for performers.

The wall surfaces of painted rough concrete and the diffusive shaping were all concealed behind a 3-foot by 3-foot grid of vertical wood dowels. Within each square of the grid, the spacing of the dowels gradually increases and then decreases to avoid a "picket fence" effect. New box seating gave a sense of human scale to the room and broke up the sheer side walls.

To serve instrumental music, the existing short, thin shell with a horizontal ceiling was replaced with a heavy wood shell as tall as the proscenium, with a projectively angled ceiling. The shell's walls are stabilized by triangular jack braces. The wall panels are tied back to the stagehouse wall for storage with the jack braces folded flat to minimize any loss of footprint in the stagehouse.

The hall was remade without significant change to its seat count. Its acoustics were greatly improved – mellower and clearer, with more presence and more reverberation, especially for orchestral performances. The reverberation time data were measured in the unoccupied auditorium with the orchestra shell in place and the few adjustable acoustic curtains in the attic volume retracted.

Opposite page: Corbett Auditorium - Interior | Kirkegaard

1. AUDIO PRODUCTION 2. STAGE 3. AUDIO CONTROL ROOM 4. TV PRODUCTION STUDIO

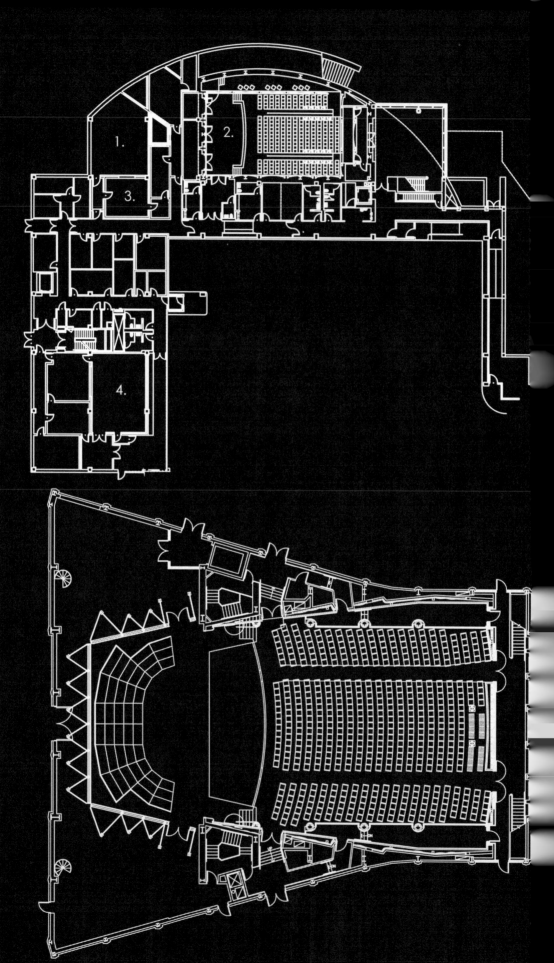

CORBETT AUDITORIUM – LONGITUDINAL SECTION

CORBETT AUDITORIUM – TRANSVERSE SECTION

ACOUSTICAL CONSULTANT:
KIRKEGAARD

ARCHITECT:
PEI COBB FREED & PARTNERS
NBBJ-ROTH

COMPLETION DATE:
2000

LOCATION:
CINCINNATI, OH I USA

CONSTRUCTION TYPE:
NEW CONSTRUCTION, RENOVATION

CONSTRUCTION/RENOVATION COST:
$93,200,000

FEATURED SPACE DATA:

ROOM VOLUME:
317,300 FT3

FLOOR PLAN AREA:
5,760 FT2

SEATING CAPACITY:
750

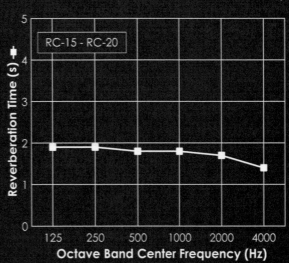

RC-15 - RC-20

Located in the arts precinct in the heart of Melbourne, the Ian Potter Southbank Centre opened in early 2019. It is a purpose-built music conservatorium and the centerpiece of the University of Melbourne's Southbank campus. Carrying a small footprint, it is organized vertically as a compact eight-level tower and aims to balance the individual concentration required of students with the camaraderie of engaging with others.

The Ian Potter Southbank Centre encourages curiosity and interaction between the musicians and their community. The circulation spaces incorporate study hubs with informal seating areas to encourage interaction and flexible use of the space. Large windows with acoustic glazing create visual connections between main spaces while maintaining acoustic isolation. The venue aims to encourage an acoustical environment that is alive with energy and that allows the sounds of learning and practice to permeate throughout the common spaces. The interior spaces are always close to natural light and views. Its stairs are natural settings for conversation. The internal ambiance is designed so that the activity of the building is audible, but not intrusive, to contribute to a supportive environment.

Each room has specialized fit-for-purpose or flexible features designed to control key acoustic measures. Ceiling heights have been specified to create room volumes which reduce sound level exposure while retaining a suitably lively acoustic impression. The design allows for flexibility in the life of the building with modular acoustic treatments that allow for easy changes as the functions of the rooms are evolving. The building includes three vertically stacked venues:

- Kenneth Myer Auditorium is a flat-floor venue for rehearsals and workshops with large orchestras and choirs. A 200-seat retractable seating bank can accomodate audiences for performances, public recitals, and guest lectures.

- Hanson Dyer Hall is a formal recital hall that holds up to 40 musicians on stage, a choral balcony that holds up to 60 musicians, and raked seating for an audience of up to 400. The seating layout is designed to feel intimate for a single performer or lecturer with a small audience or for a full theatre.

- Prudence Myer Studio is a flexible flat-floor performance and rehearsal venue, accommodating up to 135 people. This space is for performance classes and workshops, academic teaching, exams, recitals, and small concerts.

These spaces achieve a high level of acoustic isolation from the external environment, including trams along Sturt Street, by utilizing box-in-box construction. Vibration-isolated structures were built within a structurally massive concrete box and provide high-quality recording environments for the three large performance spaces. These main rooms have operable acoustic banners to change the reverberation time in the spaces to suit a wide range of functions and uses.

Kenneth Myer Auditorium demonstrates the design philosophy that emphasizes the flexibility of the building. The retractable seating allows for a recital audience of up to 200. A fixed control position linked to sound, projection, and lighting systems allows for multimedia performances and recording capability. Operable banners allow the acoustical transformation of the space for guest lectures and amplified electronic performances. A viewing gallery at upper level allows observation of rehearsals and events by an additional 30 people or a site for off-stage performers. External light is provided through niche skylights embedded in the façade, and visual interaction is achieved through external portals and internal feature windows to the circulation spaces.

With the seats retracted, the floor area extends to accommodate an orchestra of 120 and a chorus. Flexibility in the acoustical design extends to the double curvature of the ceiling panels which allows multiple overhead reflections between all areas of the floor area. Low-level wall diffusion allows the musicians to be immersed in an enveloping sound while avoiding strident reflections that could distort the sound of the ensemble. This room uses operable absorption, and the reverberation time can be varied from 1.1 to 1.3 seconds depending on the use. The background noise levels meet the NR-20 design target for the unoccupied space.

The building incorporates twelve tutorial rooms

designed for rehearsals of 12-20 musicians and a percussion space. Room heights up to 4.5 meters reduce the sound intensity for occupants. Smaller studios provide staff accommodation, as well as practice and rehearsal spaces for up to three musicians. Three electronic music studios are linked to the central recording studio to provide a comprehensive capability for demonstration and commercial-level productions.

Hanson Dyer Hall cantilevers dramatically over the park below to frame an external space, which is the fourth performance space. Here is a place outside the walls of the conservatorium for both impromptu and organized performance. It draws music into the public realm, creating a social space owned by the students. The reverberation time data presented were calculated for the unoccupied space.

Traditionally music education requires long periods of intense private practice to learn the complex skills and knowledge needed to be a musician. It can be quite a solitary pursuit. The acoustic isolation required between rehearsal spaces can reinforce the isolation of a musician from the world around them. However, the experience of learning is enhanced by encouraging students to interact with their peers, to discuss and debate ideas, and to learn from each other. This building celebrates the activities of rehearsal and practice - it is the 'green room' of Melbourne's arts precinct. Performances will take place within the building, yet it is not a concert hall. The working interior is designed specifically to support the daily life of the conservatorium - robust, informal, and playful.

In addition to the architect and acoustical consultant, the design team included: Brian Hall & Marshall Day Entertech (theatre consultant), CHW Consulting (audio systems designer, video systems designer), Electrolight (lighting designer), Aurecon (mechanical engineer), Lendlease (general contractor), Aspect Studios (landscape architect), and IrwinConsult (structural & civil engineer).

Kenneth Myer Auditorium | Trevor Mein

Hanson Dyer Hall - Lecture & Performance Space | Trevor Mein

Level 8 Tutorial Room | Trevor Mein

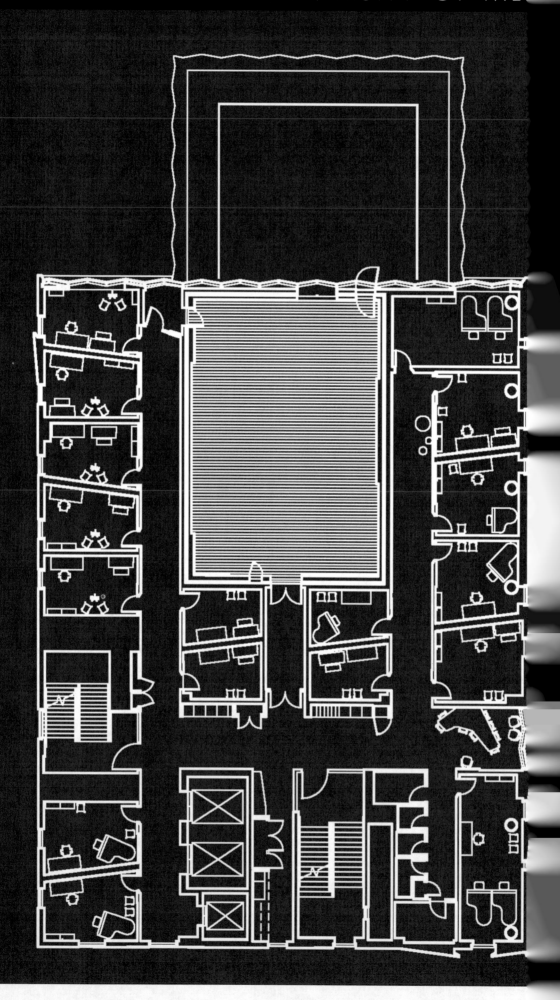

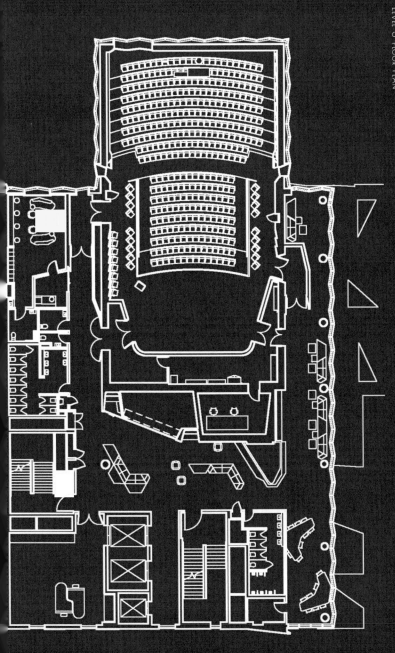

Level 3 Floor Plan

Acoustical Consultant:
Marshall Day Acoustics

Architect:
John Wardle Architects

Completion Date:
2019

Location:
Southbank, Victoria | Australia

Construction Type:
New Construction

Construction/Renovation Cost:
$70,640,433

Featured Space Data:

Room Volume:
123,600 ft3

Floor Plan Area:
5,060 ft2

Seating Capacity:
358

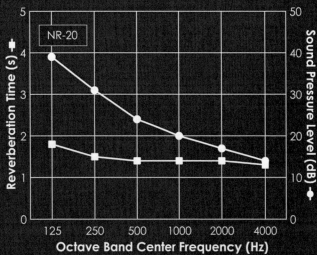

Wheaton College has long cultivated young musicians in a tight-knit community outside of Chicago, Illinois. As the program has grown in recent decades, however, the community has burst at the seams in need of expanded space for faculty and students alike. The construction of a new science building on campus led to the availability of Armerding Hall to be redeveloped into a new home for the Conservatory on the central campus quad. Acoustic isolation strategies were developed to allow for practice and instruction sessions to happen throughout the building without distractions, with technical systems supporting advanced instruction and recording capabilities on every floor.

The first phase of work completed in 2017 includes the renovation of the existing science building to include a recording studio, music technology lab, keyboarding lab, practice rooms, teaching studios, ensemble rooms, and faculty and administrative offices. An existing lecture hall has been reconfigured into a new 100-seat recital hall that has become a favored spot for classes and recitals alike.

Sound isolation was an important consideration in the design of the entire project. Limited structural capacity, floor-to-floor dimensions, and the desire to maximize room volume all provided unique opportunities for the isolation concept. Separate stud multi-drywall partitions were the primary means of acoustic separation in order to keep structural loads at a minimum. Key spaces including the ensemble rooms and recording rooms were able to

receive masonry construction in addition to drywall partition build-ups. Resilient ceiling construction was strategically shaped in many spaces around building services and held tight to structure in order to maximize room volume.

The acoustic goal for the 100-seat recital hall was to create an intimate room with a pleasing reverberation for music with the ability to successfully facilitate lectures. The limited room volume of the former lecture hall was expanded by removal of the former steeply raked seating to capture more room volume. Further strategies to sustain reverberation involved the angling of side walls in section to encourage additional late energy development. The existing rooms masonry walls were taken of advantage as a part of the room envelope composition in order to

support low frequency sound while textured GFRG panels laminated to the upper side walls played a key role in tempering high frequency energy while sustaining longer reverberation times.

Acoustic variability plays a key role in the space, realized by upper side wall drapery and lower upstage wall drapery. In the unoccupied room the reverberation has a swing of nearly 1 second at mid frequencies. The reverberation time data shown were measured in the occupied space. The room is able to adapt to support music rehearsal and performance, spoken word, and light amplified sound favorably. The background noise rating shown was calculated for the unoccupied space with the HVAC system on.

A second phase of the project will include an addition of a 650-seat concert hall and the college's first dedicated rehearsal hall for their formidable choral program.

In addition to the architect and acoustical consultant, the design team included: Theatre Projects Consultants (theatre consultant, lighting designer), Threshold Acoustics LLC (audio systems designer), IMEG Corp (mechanical engineer), and Mortenson Company (general contractor).

Opposite page: Recital Hall | Darris Lee Harris

FACULTY STUDIO | DARRIS LEE HARRIS

EXTERIOR VIEW RENDERING | HGA ARCHITECTS AND ENGINEERS

1. Concert Hall 2. Lobby 3. Choral Rehearsal 4. Classrooms 5. Dressing Rooms 6. Office 7. Recital Hall 8. Storage 9. Lounge 10. Hospitality

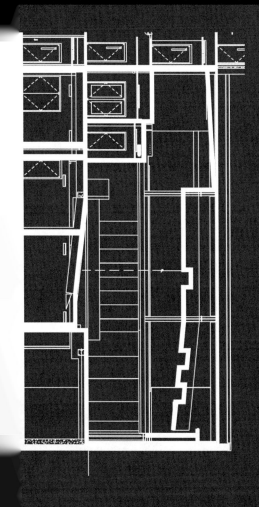

RECITAL HALL – LONGITUDINAL SECTION

RECITAL HALL – PLAN

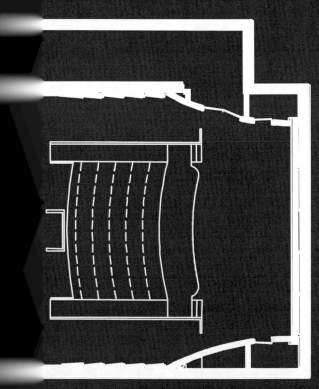

ACOUSTICAL CONSULTANT:
THRESHOLD ACOUSTICS LLC

ARCHITECT:
HGA / FGM

COMPLETION DATE:
2017

LOCATION:
WHEATON, IL I USA

CONSTRUCTION TYPE:
ADAPTIVE REUSE, FUTURE EXPANSION

CONSTRUCTION/RENOVATION COST:
$38,000,000 [$17 MILLION
(PHASE 1), $21 MILLION (PHASE 2)]

FEATURED SPACE DATA:

ROOM VOLUME:
56,000 FT3

FLOOR PLAN AREA:
2,100 FT2

SEATING CAPACITY:
100

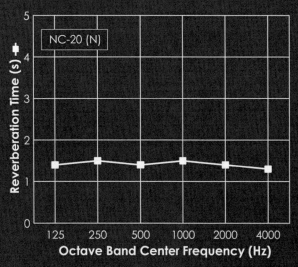

Founded in 2003, the Yong Siew Toh Conservatory of Music in Singapore has quickly established a reputation as an exciting international conservatory and one of the world's most distinctive institutions for training and educating performers, composers, and recording engineers. The Yong Siew Toh Conservatory of Music building opened in 2006 and was designed by RSP Architects Planners and Engineers.

The conservatory pays homage to the environment's tropical brightness through the use of glass panels, and on the main façade, one can look through the full-length glass panels into the foyer of the concert hall and into the library. Despite the grandeur of the building as it occupies 26,000 square meters, its spaces provide intimate areas for conversations and interactions.

The Institution offers majors such as instrumental performance, voice, composition audio arts and science, and music and society. The facilities include a 650-seat concert hall, an orchestral hall, recital studios, seminar rooms, eight ensemble rehearsal rooms, forty teaching studios, fifty practice rooms, a library, and a large recording studio.

Kahle Acoustics (Belgium) and Theatre Projects (UK) collaborated on acoustics and theater planning, in particular on the design of the 600-seat concert hall, while taking into account the specific context of the site and its surroundings. As the conservatory was built just off the Ayer Rajah Expressway, full sound isolation in all performance and teaching areas had to be considered. Consequently, from the smallest

Concert Hall | Yong Siew Toh Conservatory of Music

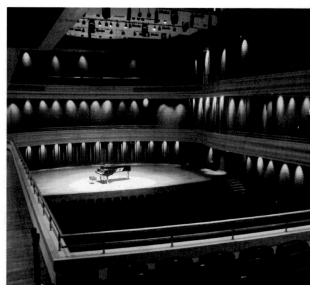

Concert Hall – Stage | Kahle Acoustics

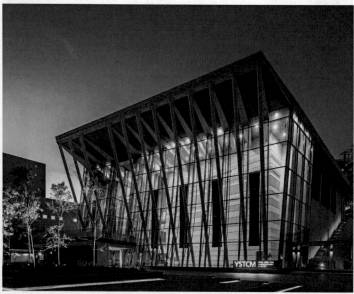

Glass-paneled Façade | Yong Siew Toh Conservatory

Orchestra Rehearsal | Yong Siew Toh Conservatory

ACOUSTIC REFLECTOR PANELS | KAHLE ACOUSTICS

ADJUSTABLE CIRCULAR ACOUSTIC CANOPIES | KAHLE ACOUSTICS

CURVED BALCONIES WITH MAPLEWOOD FINISH | KAHLE ACOUSTICS

rehearsal rooms to the main concert hall, each acoustically sensitive space was built as a box within a box. In some cases, the wedge-shaped design was applied in order to prevent flutter echoes.

The 600-seat concert hall was built for classical music performances, chamber music, and symphony orchestra concerts. Its rectangular shoebox shape was based on the designs of other highly-regarded concert halls including the Amsterdam Concertgebouw, the Musikverein in Vienna, and the Boston Symphony Hall. Its seating area has a narrow width that creates a feeling of intimacy. Sound reflectors above the stage with adjustable heights contribute to the variable acoustics; a higher setting of the canopy is utilized when there is a need for a larger acoustic volume. The surface finishes have both architectural value and acoustic properties. The balcony fronts are finished in maple wood, and their curved surfaces of various depths help distribute the sound throughout the hall. Grooved surfaces of varying depths diffuse sound waves to every corner of the room. The wavy finishes of the wall panels placed in front of the acoustic curtains serve as storage pockets but also create a similar diffusion effect. The reverberation time data were measured in the unoccupied space, and the noise ratings were calculated for the unoccupied space.

The Yong Siew Toh Conservatory of Music of Singapore hosts 200 concerts annually, featuring students and faculty staff alongside international artists and ensembles as a commitment to local, regional, and global outreach. The Conservatory strongly contributes to a distinct and powerful contemporary Asian voice that is both collaborative and productive in an ever-flamboyant artistic landscape. Overall the acoustics support what the conservatory stands for, a place that "nurtures artistic identity as a driver and compass for excellence."

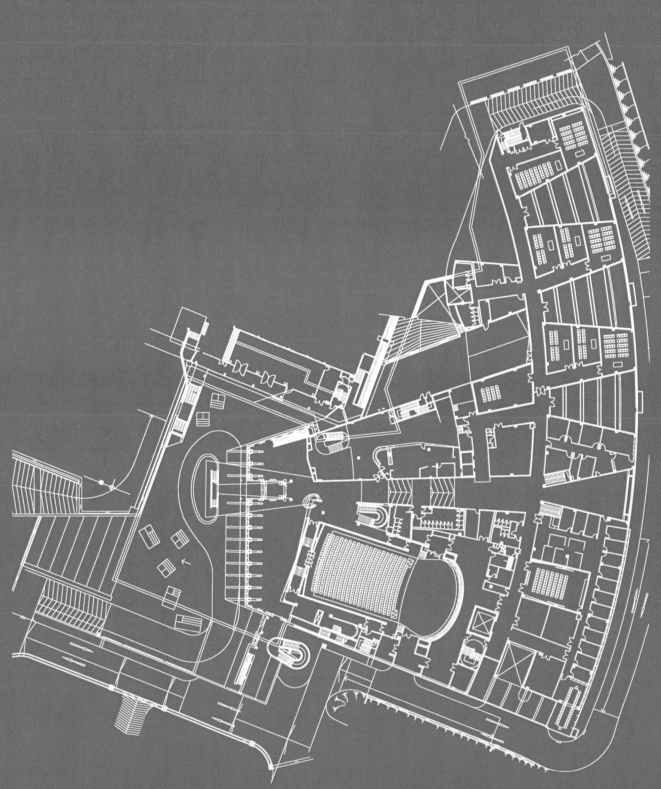

ACOUSTICAL CONSULTANT:
KAHLE ACOUSTICS

ARCHITECT:
LIU THAI KER, RSP ARCHITECTS

COMPLETION DATE:
2006

LOCATION:
SINGAPORE

CONSTRUCTION TYPE:
NEW CONSTRUCTION

CONSTRUCTION/RENOVATION COST:
$21,718,650

FEATURED SPACE DATA:

ROOM VOLUME:
317,832 FT3

FLOOR PLAN AREA:
6,889 FT2

SEATING CAPACITY:
650

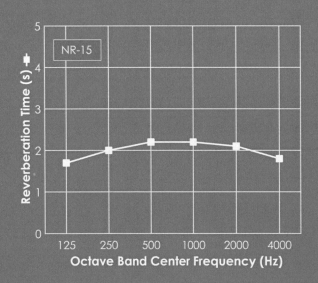

APPENDIX A
GLOSSARY

absorbing materials: materials that dissipate acoustic energy within their structure as heat and/or mechanical vibration energy. Usually, building materials designed specifically for the purpose of absorbing acoustic energy on the boundary surfaces of rooms or in the cavities of structures [1].

absorption: the process of dissipation of sound energy, or the sound energy absorbed by a component or an entire room [2].

absorption coefficient (α): a measure of the sound absorbing property of a surface. More specifically, absorption coefficient is defined as the fraction of the incident sound energy absorbed (or otherwise not reflected) by a surface [2].

acoustical transparency: a measure of the amount of sound that passes through an opening or more commonly through a screen or sound-absorbing panel. The amount of sound that passes through is dependent on the size and distribution of the voids in the screen as a result of how sound bends as it travels through openings [2]. See also *diffraction.*

acoustical treatment: the use of acoustical absorbing, reflecting, diffusing, or transmissing materials or sound-isolating structures to improve or modify the acoustical environment [1].

acoustics: the science of sound, including its production, transmission, and reception [3].

acoustic (sound) environment: the overall environment, interior to exterior, which affects the acoustic conditions of the space or structure under consideration [1].

airborne sound: sound transmitted through air as a medium rather than through solids, such as the structure of a building [1].

air-handling unit (AHU): an assembly of air-conditioning components (cooling coils, filters, fan, humidifiers, dampers, etc.) installed in a room or assembled into a self-contained unit. An AHU delivers air to an air distribution (duct) system [2].

ambient noise: See *background noise.*

amplification: usually, the increase in intensity level of an audible signal produced by means of loudspeakers and associated electrical amplification apparatus [1].

arena theatre: theatres that have an audience around four sides of the stage. Also known as amphitheatres, island stage theatres, center stage theatres, or theatre-in-the-round [4].

articulation index (AI): a calculated coefficient used for rating the intelligibility of speech [3]. A value of zero implies zero speech intelligibility, and a value of 1, a perfect intelligibility [2].

articulation loss of consonants (ALCons): a calculated value used for rating the intelligibility of speech that is typically expressed as a percentage (%ALCons). Lower percentages imply less articulation information lost, or better speech intelligibility. Higher percentages imply more articulation information lost, or worse speech intelligibility [5].

attenuation: reducing the magnitude of a sound signal by separation of a sound source from a receiver, acoustical absorption, enclosure, active cancellation by electronic means or a combination of these or other means [1].

audible: capable of producing the sensation of hearing [1].

auditorium: large event space reserved for special programs and gatherings.

A-weighting: the standard frequency weighting that de-emphasizes low frequency sound similar to average human hearing response and approximates loudness and annoyance of noise. A-weighted sound levels are frequently reported as dBA [3].

background noise: the generally lowest or residual sound level present in a space above which speech, music, or other sounds may be heard [1]. Typically measured while the space is unoccupied and may include airborne, structure-borne, or equipment noise [2]. Common sources are a building's mechanical, electrical, plumbing, fire protection, or native building system, outside noise such as traffic, etc. Also called ambient noise.

backstage: those elements of a theatre that serve primarily the aesthetics of theatre performance. Backstage facilities can include the stage support spaces such as wing space, trap rooms, fly spaces, and rear stages, all of which also allow for multiple entrances and exits for the actors. Backstage spaces can also include dressing rooms, green rooms, rehearsal rooms, production spaces such as shops for costumes and scenery, laundry facilities, loading docks, security stations, storage areas, and control rooms for stage lighting, sound, and special effects [4].

balanced noise criteria (NCB): a numerical rating system, or family of curves used to specify background sound levels over a given frequency range. NCB is a revision of the NC method, with alterations including extension into the low frequencies, calculation of the speech interference level (SIL), and ratings of spectral quality [5]. See also *Noise Criteria and speech interference level.*

balcony seating: seating areas raised above the main floor to offer better sight lines between the audience and the performance area. Also referred to as gallery seating.

black box theatre: See *multi-form theatre.*

brilliant: a bright, clear, ringing sound, rich in harmonics. A brilliant sound has prominent treble frequencies that decay slowly [6].

broadband noise: a sound whose sound pressure level plot has no dominant peaks, varying smoothly with frequency [2].

caliper: ramps leading from the audience to the stage.

chamber music: music written for small rooms or salons that is performed by a small group of musicians [5].

clarity (C50): the clarity factor. It is the ratio, expressed in decibels, of the energy of the first 50 milliseconds of an impulse sound arriving at the listener's position divided by the energy in the sound after 50 milliseconds. The divisor is approximately the total energy of the reverberant sound [6].

coffered ceiling: a ceiling design in which upward indentations are framed by beams arranged in a grid pattern [7].

comb filtering: an effect resulting from an alternating sequence of constructive and destructive interference of sound waves across frequency. In other words, the sound waves either add together to form a higher sound level or cancel each other out depending on the frequency. The resultant signal displays a series of dips, shaped much like the teeth of a comb. This effect is more commonly found in small studios [5].

concert hall: theatre built solely for music [8].

cornice: the decorated projection at the top of a wall provided to protect the wall face or to ornament and finish the eaves. The term is used as well for any projecting element that crowns an architectural feature, such as a doorway [9].

curtain wall: a thin, usually aluminum-framed wall, containing in-fills of glass, metal panels, or thin stone. The framing is attached to the building structure and does not carry the floor or roof loads of the building [10].

dBA: A-weighted sound pressure level. See also *A-weighting.*

dead room: a room that is characterized by a large amount of sound-absorbing material and with little reverberation [3]. Also known as a "dry" room. See also *live room.*

decibel (dB): the measurement unit used in acoustics for expressing the logarithmic ratio of two sound pressures, intensities, or powers. Typically used to describe the magnitude of a sound with respect to a reference level equal to the threshold of human hearing [3].

diffraction: a change in the direction of propagation of sound as a result of bending caused by a barrier in the path of the sound wave [2].

diffusion: dispersion of sound within an enclosure such that there is uniform energy density throughout the space [1]. With diffusion in the room, the listener receives sound from various directions [2].

direct sound: the sound that arrives at a receiver along a direct line from the source without reflection or other interaction from any surface [2].

distribution: the pattern of sound intensity level within a space; also, the patterns of sound dispersion the sound travels within a space [1].

dynamic range: the range in decibels between the maximum and minimum signal levels from a device or sound source [1].

early decay time (EDT): measured with same technique as reverberation time with the following modification - it is the time it takes for a signal to decay from 0 to -10 dB relative to its steady state value. A multiplying factor of 6 is necessary to make EDT time comparable to reverberation time [6]. See also *reverberation time.*

echo: a sound that has been reflected with sufficient time delay, and is of a sufficiently high level to be heard as distinct from the original sound [2].

ensemble: a group of musicians.

equivalent (average) sound pressure level (Leq): the average sound pressure level occurring in a specified period (e.g., 1 hour) [3]. More specifically, it represents the sound pressure level which, if constant over a given period, will contain the same sound energy as the actual sound that is fluctuating with time over that period [2].

façade: the front of a building. Can also refer to any face of a building given special architectural treatment [11].

fenestration: Originally, an architectural term for the arrangement of windows, doors and other glazed areas in a wall. Has evolved to become a standard industry term for windows, doors, skylights and other glazed building openings [12].

floating floor: an additional layer of floor (concrete or wood) supported on a structural floor (concrete or wood) through resilient mounts. To be effective, the floating floor must be isolated at all sides from walls or other building components, so that the impact or vibration from the floor does not flank to other parts of the building through the wall [2].

flutter echo: a rapid, but repetitive succession of sounds from a sound source, usually occurring as a result of multiple reflections in a space with reflective, smooth, and parallel walls [2].

focus: a concentration of reflected acoustic energy within a limited location in a room, most often caused by room geometry, such as concave surfaces [1].

forestage: the portion of stage that exists in front of the proscenium or curtain line [1]. Also known as apron.

frequency: the number of (full) cycles per second measured. The unit of frequency is cycles per second (cps), which is also called Hertz (Hz). A frequency of 500 Hz means 500 cycles per second [2].

frequency band: a subdivision of the frequency range of interest for measurement or analysis purposes (e.g., octave band, one-third-octave band, etc.) [1].

glazing: the glass (and other materials) in a window or door. Also, the act or process of fitting a unit with glass [12].

Haas effect: See *precedence effect.*

Hertz: Unit of measurement of frequency. See also *frequency.*

house: audience seating area of a theatre.

HVAC: heating, ventilation, and air-conditioning systems which condition the spaces for occupant comfort.

impact insulation class (IIC): a single-number rating derived from measured values of impact sound pressure levels, which provides an estimate of the impact sound insulating performance of a floor-ceiling assembly [2].

initial-time-delay gap (ITDG): the interval in milliseconds between the arrival of the direct sound and the first reflection at the listener. The smaller this interval, the more intimate the hall, or the more involved the listener feels with the performance [2].

intimacy: the subjective impression of the size of the hall. A small hall helps establish a sense of closeness to the source, giving a feeling of intimacy. The opposite feeling, usually obtained in wide halls, is one of being "detached" or "remote" from the performance. The acoustical measure of intimacy is called the initial time delay gap [2]. See also *initial time delay gap (ITDG).*

lite: a piece of glass. In windows and doors, lite refers to separately framed panes of glass (as well as designs simulating the look of separately framed pieces of glass) [12].

live room: a room containing a relatively small amount of sound absorption, leaving a space with a longer reverberation time. A reverberation chamber is an extremely live room [2]. See also *dead room as an antonym.*

loudness: an auditory sensation that varies with sound pressure level and is also dependent on both frequency and time [3].

mullion: a component used to structurally join two window or door units [12].

multi-form theatre: theatres that do not establish a fixed relationship between the stage and the house.

They can be arranged into any of the standard theatre forms or any of the variations of those. Also known as black box theatres, laboratory theatres, flexible stage theatres, modular theatres, free form theatres, or environmental theatres, they can be reconfigured for each performance [4].

noise: unwanted sound [2].

noise (sound) control: the application of acoustical principles to the design of structures, equipment, and spaces to permit them to function properly and to create the desired environment for the activities intended [1].

noise criteria (NC): a numerical rating system, or family of curves used to specify background sound levels over a given frequency range [8].

noise rating (NR): a numerical rating system based on a set of octave band curves used primarily in Europe and the United Kingdom to describe noise from mechanical ventilation systems in buildings [5].

noise reduction: the reduction in level of unwanted sound by any of several means (e.g., by distance in outdoor space, by boundary surface absorption, by isolating barriers or enclosures, etc.) [1]. The term is also used to describe the difference in sound pressure levels between two rooms, in which one room is the source room and the other, a receiving room [2].

noise reduction coefficient (NRC): a single-number rating derived from measured values of sound absorption coefficients of a material at 250, 500, 1000, and 2000 Hz. NRC is an estimate of the sound absorptive property of an acoustical material [2]. NRC values range from 0 (hard reflective materials) to 1.0 (highly absorptive materials).

octave (or octave band): interval between two frequency limits whose frequencies are related to each other in the ratio of 1:2. In other words, the upper frequency limit of an octave band is twice its lower frequency limit. An octave band is specified by its center frequency [2]. The standard acoustical octave bands are centered at 16, 31.5, 63, 125, 250, 500, 1000, 2000, 4000, and 8000 Hz [1].

one-third octave band: an octave band divided into three bands. The upper frequency limit of a one-third octave band is $2^{1/3}$ times its lower frequency limit [2].

parterre seating (level): the rear section of seats, and sometimes also the side sections of the main floor of a theatre, concert hall, or opera house [13].

pitch: See *frequency.*

plenum: an air-filled space in a structure, especially one that receives air from a blower for distribution (as in a ventilation system) [8]. They are typically made of sheet metal with fiberglass lining the walls [2].

plinth: lowest part, or foot, of a pedestal, podium, or architrave (molding around a door). It can also refer to the bottom support of a piece of furniture or the usually projecting stone coursing that forms a platform for a building [9].

precedence effect: the ability of the human auditory system to suppress perception of echoes, primarily up to 40 milliseconds after a direct sound [3].

preferred noise criteria (PNC): a numerical rating system, or family of curves used to specify background sound levels over a given frequency range. PNC is a revision of the NC curves, with alterations in the high- and low- frequencies [5]. See also *Noise Criteria.*

proscenium theatre: a variety of end stage theatre that intentionally puts the stage in a separate volume of space from that occupied by the house. In this form, the stage is separated from the house by a large arched opening (the proscenium, which can sometimes be rectangular or square) that allows the audience to see through from the house to the stage as if looking through a frame at a large moving picture. Proscenium stages can be fitted with "aprons" or "forestages" that carry the stage through the arch into the house and thus make it a variation on the end stage form [4].

psychoacoustics: the scientific study of human auditory perception [3].

quadratic residue diffuser: a type of diffuser that consists of an array of linear slits (or wells) of constant width and varying depth. The exact design (e.g., number of wells, size of wells, etc.) can be tuned to provide sound scattering within any required frequency band [2].

recital hall: theatre built for solo or small chamber performances of instrumentalists or vocalists.

reflected sound: the resultant sound energy returned from a surface(s) that is not absorbed or otherwise dissipated upon contact with the surface(s) [1].

reflection: the phenomenon by which a sound wave is re-radiated ("bounced") from a surface [3].

resilient mounting: any mounting or attachment system that reduces the transmission of vibrational energy from a vibrating body or structure to an adjacent structure [1].

reverberant (or reflected) sound field: a sound field created by repeated reflections of sound from the boundaries of an enclosed space [2].

reverberation: the persistence of a sound within a space after the sound source has stopped [1].

reverberation time (RT): the time (in seconds) required for the sound pressure level to decrease 60 dB in a room after a noise source is abruptly stopped. Reverberation time relates to a room's volume and sound absorption [3].

room acoustics: a study of the behavior of sound in rooms [2] that focuses on the characterization and optimization of the acoustical conditions within a built environment.

room criteria (RC): a numerical rating system, or family of curves used to specify background sound levels over a given frequency range. Includes calculation of a level rating based on the arithmetic mean of the noise levels at 500, 1000, and 2,000 Hz and ratings of spectral quality [5].

room shape: the configuration of an enclosed space, resulting from the orientation and arrangement of surfaces defining the space [1].

sabin: a unit of measure of sound absorption. The unit "sabin" can be either foot sabin or metric sabin, depending on whether the surface area of the absorber is measured in ft^2 or m^2. One foot sabin is the sound absorption provided by 1 square foot of a surface whose absorption coefficient is 1.0. Similarly, 1 metric sabin is the sound absorption provided by 1 square meter of a surface whose absorption coefficient is 1.0 [2]. Named after Wallace Clement Sabine, a pioneer in architectural acoustics [2]. See also *absorption and absorption coefficient (α)*.

scattering: an irregular reflection and/or diffraction of sound in many directions [2].

sectional: practice session for a specific group of an ensemble such as brass, woodwinds, strings, or specific instruments.

shoebox hall: rectangular hall that is relatively narrow. Floors are generally flat with the orchestra seated above the heads of the patrons on a high, raked platform [5].

signal-to-noise ratio (S/N): the difference in dB between the signal (e.g., speech) level and the noise level [2].

sound isolation: a lack of acoustical connection. There are generally two ways of achieving sound isolation: (i) by insulation, that is, by inserting a barrier between the source and the receiver, and (ii) by attenuation, that is, by reducing the intensity of sound as the sound travels from the source to the receiver [2].

sound lock: in architectural acoustics, a small space that works as a buffer between a source room and a receiving room. A sound lock is usually connected to the source and receiving rooms with a series of acoustically gasketed doors [2].

sound pressure level (SPL): sound pressure measured as a level in decibels typically referenced to 20 micro-Pascals , generally at a specific location or distance from a sound source [3].

sound transmission: the propagation of sound energy through various media [1].

sound transmission class (STC): a single-number rating derived from laboratory measurement of sound transmission loss. The STC describes the sound-insulating properties in the 100 – 4000 Hz frequency range [3].

sound transmission loss (TL): a laboratory measure of sound insulation indicative of the sound intensity flow transmitted through a partition without regard to the partition size, usually measured in one-third octave bands [3].

sound wave: sound travels in space in the form of sound waves, which is similar to the motion of a ripple produced by dropping a pebble into a pond of water [2].

specular reflection: a mirror type reflection, similar to the reflection of light from a mirror. The reflected sound path makes the same angle with the reflecting surface as the incident sound path [2].

speech intelligibility: a measure of how well speech can be understood by a listener [2].

speech interference level (SIL): a single-number rating used to evaluate interference based on the background noise level and voice level [3]. The SIL is the arithmetic mean of the noise levels at 500, 1000, 2000 and 4000 Hz [5].

speech transmission index (STI): a calculated coefficient used for rating the intelligibility of speech [5]. A value of zero implies zero speech intelligibility, and a value of 1, a perfect intelligibility.

structure-borne sound: sound energy transmitted through solid elements of a building structure [2].

thrust stage theatre: theatres in which the stage thrusts out from one side of the space into the midst of the audience. The audience is most often located around three sides of a thrust stage, though they can be located on two sides opposite each other. Also known as open stage theatres and sometimes as courtyard theatres. Often, arena theatres are designed for easy conversion into thrust stages theatres by way of the removal of one section of audience seating [4].

transmission: See *sound transmission.*

transmission loss (TL): See *sound transmission loss.*

vestibule: an entrance hall inside a building [14].

vibration: a periodic motion of molecules in an elastic medium with respect to equilibrium. Vibration by mechanical equipment can be a factor in structure-borne noise radiation [3].

vibration isolation: the methods used to reduce vibration in a structure caused by vibrating equipment, including the use of springs and elastomeric materials [3].

warmth: a perception of darker sounds with richer bass resulting from a slight increase in low-frequency reverberation to mid-to-high-frequency reverberation [2].

wavelength (λ): the distance between adjacent regions of a sound wave where identical conditions of particle displacement or pressure occur [1].

Note: some definitions are provided verbatim from the cited text. Please refer to the texts cited for additional explanations or clarifications.

References

[1] Cavanaugh, William J. and Joseph A. Wilkes. *Architectural Acoustics: Principles and Practice.* John Wiley & Sons, Inc., 1999, New York.

[2] Mehta, Madan, Johnson, Jim, and Rocafort, Jorge. *Architectural Acoustics: Principles and Design.* Prentice-Hall, Inc., 1999, New Jersey.

[3] Salter, Charles M. *Acoustics: Architecture, Engineering, and the Environment.* William Stout Publishers, 1998, California.

[4] "theatre design." *Encyclopedia Britannica.* Online. 13 May 2010 from www.britannica.com.

[5] Long, Marshall. *Architectural Acoustics.* Elsevier Academic Press, 2006, Massachusetts.

[6] Beranek, Leo L. *Concert Halls and Opera Houses: Music, Acoustics, and Architecture.* Springer, 2004, New York.

[7] Howard, Hugh. "All You Need to Know About Coffered Ceilings." *Bob Vila Tried, True, Trustworthy Home Advice,* www.bobvila.com/articles/coffered-ceilings/. Accessed 19 December 2019.

[8] Hardy, Hugh and Kliment, Stephen A. *Building Type Basics for Performing Arts Facilities.* John Wiley & Sons, Inc., 2006, New York.

[9] Encyclopaedia Britannica. Online. Accessed 19 December 2019 from www.britannica.com.

[10] Vigener, Nik. "Curtain walls." *Whole Building Design Guide,* Revised by Richard Keleher and Rob Kistler, 10 May 2016. www.wbdg.org/guides-specifications/building-envelope-design/fenestration-systems/curtain-walls. Accessed 19 December 2019.

[11] Merriam-Webster Online Dictionary. Accessed 19 December 2019 from www.merriam-webster.com/dictionary.

[12] "Glossary of Terms: Window & Door Terms You Need to Know." *Windsor Windows and Doors A Woodgrain Millwork Company,* www.windsorwindows.com/knowledge-center/windows-and-doors-101/glossary. Accessed 19 December 2019.

[13] Dictionary.com. Accessed 19 December 2019 from www.dictionary.com.

[14] Merriam-Webster Online Dictionary. Accessed March 2015 from www.merriam-webster.com/dictionary.

APPENDIX B:
NOTES ON CURRENCY, UNITS, AND SCALE

All reported construction and renovation costs are approximate. In some cases, it is difficult to single out the cost of an individual music education facility when it is part of a larger complex. The costs given in this publication were those reported by the contributing consulting firm, and questions regarding specific costs should be directed to the individual firms.

All currencies have been converted to United States Dollars (USD) based on the exchange rate for each currency at the time of publication. The units for room dimensions in the information bar at the end of each music education facility spread are shown in English units. However, the units given in individual music education facility descriptions were provided by the contributing consulting firm, as stated in the description.

The architectural drawings of the music education facilities have not been uniformly scaled, and therefore should not be compared directly without modifying the drawing scale. Instead the drawings are shown in the largest size possible, which varies for each page spread. This format was chosen over uniformly scaled drawings in favor of showing more architectural detail for each space. Relative scale for the drawings can be determined by measuring off common architectural features such as standard doorways. Overall, these images have been provided to give the reader a sense of the layout of the spaces, and should not be read as official architectural documents.

L. M. Ronsse et al. (eds.), *Rooms for the Learned Musician*

APPENDIX C:
REFERENCES

Beranek, Leo L. "Concert Hall Acoustics-2008." *Journal of the Audio Engineering Society*, vol. 56, no. 7/8, 2008, pp. 532-544.

---. *Concert Halls and Opera Houses: Music, Acoustics, and Architecture*. Springer, 2004, New York.

---. *Noise and Vibration Control*. McGraw-Hill, 1971, New York.

Boner, Charles K. and Coffeen, Robert C. *Acoustics for Performance, Rehearsal, and Practice Facilities: A Primer for Administrators and Faculties*. National Association of Schools of Music, 2000, Reston, Virginia.

Bradley, David T., Ryherd, Erica E., and Vigeant, Michelle C. (Eds.). *Acoustical Design of Theatres for Drama Performance: 1985 – 2010*. Acoustical Society of America, 2010, New York.

Bradley, David T., Ryherd, Erica E., and Ronsse, Lauren M. (Eds.). *Worship Space Acoustics: 3 Decades of Design*. Springer, 2016, New York.

Caller. "You Don't Build a Church for Easter Sunday: A Good Lesson for Contact Center Capacity Planning." *Contact Center 411*, 24 April 2014, http://contactcenter411.com/you-dont-build-a-church-for-easter-sunday-a-good-lesson-for-contact-center-capacity-planning/. Accessed 25 June 2020.

Carlson, Matt. "The Joke." *Carnegie Hall*, https://blog.carnegiehall.org/Explore/Articles/2020/04/10/The-Joke. Accessed 25 June 2020.

Cavanaugh, William J. and Wilkes, Joseph A. *Architectural Acoustics: Principles and Practice*. John Wiley & Sons, Inc., 1999, New York.

Dictionary.com. Accessed 19 December 2019 from www.dictionary.com.

Dobrin, Peter. "Lenfest Hall Opens a New Era for Curtis Institute of Music." *The Philadelphia Inquirer*, 6 September 2011.

Doelle, Leslie L. *Environmental Acoustics*. McGraw-Hill, 1972, New York.

Egan, M. David. *Architectural Acoustics*. McGraw-Hill, 1988, New York.

Encyclopaedia Britannica. Online. Accessed 19 December 2019 from www.britannica.com.

Eplee, David F. *An Acoustical Comparison of the Stage Environments of the Vienna Grosser Musikvereinsaal and a Scale Model*. 1989. University of Florida, Master of Architecture Thesis.

Field of Dreams. Directed by Phil Alden Robinson, Universal, 1989.

Gade, A. C. "Musicians Ideas about Room Acoustic Qualities." *Technical University of Denmark*. Report No. 31, 1981.

"Glossary of Terms: Window & Door Terms You Need to Know." *Windsor Windows and Doors A Woodgrain Millwork Company*, www.windsorwindows.com/knowledge-center/windows-and-doors-101/glossary. Accessed 19 December 2019.

Hardy, Hugh and Kliment, Stephen A. *Building Type Basics for Performing Arts Facilities*. John Wiley & Sons, Inc., 2006, New York.

Howard, Hugh. "All You Need to Know About

Coffered Ceilings." *Bob Vila Tried, True, Trustworthy Home Advice*, www.bobvila.com/articles/coffered-ceilings/. Accessed 19 December 2019.

ISO 16:1975. *Acoustics - Standard tuning frequency (Standard musical pitch)*. (Standard No. 3601). Retrieved from https://www.iso.org/standard/3601.html.

Knudsen, Vern O. and Harris, Cyril M. "Acoustical Designing in Architecture." *Published for the Acoustical Society of America by the American Institute of Physics*. 1950.

Lubman, David and Wetherill, Ewart A. (Eds.). *Acoustics of Worship Spaces*. Acoustical Society of America, 1985, New York.

Long, Marshall. *Architectural Acoustics*. Elsevier Academic Press, 2006, Massachusetts.

Martineau, Jason. *The Elements of Music: Melody, Rhythm, & Harmony*. Bloomsbury Publishing USA, 2008, New York.

McCue, Edward and Talaske, Richard H. (Eds.). *Acoustical Design of Music Education Facilities*. Acoustical Society of America, 1990, New York.

Mehta, Madan, Johnson, Jim, and Rocafort, Jorge. *Architectural Acoustics: Principles and Design*. Prentice-Hall, Inc., 1999, New Jersey.

Merriam-Webster Online Dictionary. Accessed 19 December 2019 from www.merriam-webster.com/dictionary.

Meyer, Jurgen. *Acoustics and the Performance of Music*, 5th edition, Springer Science and Business Media, 2009, New York.

Juslin, Patrick N. and Sloboda, John A. (Eds.). *Music and Emotion: Theory and Research*. Oxford University Press, 2001, New York.

Norwegian Standard NS 8178-2014. *Acoustic Criteria for Rooms and Spaces for Music Rehearsal and Performance*. Standard Norge. ICS 91.120.20.

Parkin, Peter H. and Humphreys, Henry Robert. *Acoustics, Noise and Buildings*. Praeger, 1958, New York.

Pirn, Rein. "On the Loudness of Music Rooms." *The Journal of the Acoustical Society of America*, vol. 53, no. 301, 1973, p. 301.

Sabine, Wallace Clement. *Architectural Acoustics*. Paper No. 2 in *Collected Papers on Acoustics*. Harvard University Press, 1923. Published 1964 by Dover Publications Inc., New York, p. 72-77.

Salter, Charles M. *Acoustics: Architecture, Engineering, and the Environment*. William Stout Publishers, 1998, California.

Schafer, R. Murray. *The Soundscape: Our Sonic Environment and the Tuning of the World*. Destiny Books, 1977, Rochester, Vermont.

Schröder, Manfred R. "Diffuse Sound Reflection by Maximum-Length Sequences." *The Journal of the Acoustical Society of America*, vol. 57, no. 1, 1975, pp. 149-150.

Siebein, Gary W. "An Exploration of the Urban Design Possibilities Offered by Soundscape Theory". *Proceedings of the AESOP-ACSP Joint Congress*, 2013, Dublin.

---. "Creating and Designing Soundscapes." In Kiang,

J. et al. (Eds.). *Soundscapes of European Cities and Landscapes*. COST Office, Soundscape COST-2013, Oxford.

---. *Project Design Phase Analysis Techniques for Predicting the Acoustical Qualities of Buildings*. University of Florida Architecture and Building Research Center, 1986, Gainesville, Florida.

Siebein, Gary W. and Cervone, Richard P. "Listening to Buildings: Experiencing Concepts in Architectural Acoustics." *Education Honors Monograph*, In Bilello, J. (Ed.). American Institute of Architects, 1992, Washington D.C.

Sobolewski, Rich. "The 'Temple of Music' Reopens" *Visions, Mississippi University for Women*, www.muw.edu/visions/features/120-the-temple-of-music-reopens. Accessed 20 December 2019.

Talaske, Richard H. and Boner, Richard E. *Theatres for Drama Performance: Recent Experiences in Acoustical Design*. American Institute of Physics, 1986, New York.

Poeppel, David, Overath, Tobias, Popper, Arthur N., and Fay, Richard R. (Eds.). *The Human Auditory Cortex*. Springer 2012, New York.

"theatre design." Encyclopedia Britannica. Online. 13 May 2010 from www.britannica.com.

Truax, Barry. *Acoustic Communication*. 2nd edition, Ablex Publishing, 2001, Westport, Connecticut.

Tsaih, Lucky Shin-Jyun. *Soundscape of Music Rehearsal in Band Room*. 2011. University of Florida, Ph.D. Dissertation.

Vigener, Nik. "Curtain walls." *Whole Building Design Guide*, Revised by Richard Keleher and Rob Kistler, 10 May 2016. www.wbdg.org/guides-specifications/building-envelope-design/fenestration-systems/curtain-walls. Accessed 19 December 2019.

Wenger Corporation. *Planning Guide for School Music Facilities*. 2008. www.wengercorp.com/Construct/docs/Wenger%20Planning%20Guide.pdf.

Wenger Corporation. *Planning Guides for New Construction and Renovation: Acoustics Primer*. Wenger Corporation, 2001, Minnesota.

INDEX A:
BY SPACE TYPE

Art Gallery / Art Studio 42, 62, 152, 156,
210, 179

Classroom 34, 43, 50, 58, 66, 79, 92,
100, 113, 125, 141, 152, 161, 170, 179,
187, 195, 210, 218, 226, 234, 242, 260,
280

Control Room 63, 83, 93, 157, 183, 206,
223, 231

Halls:
   Concert 42, 92, 108, 120, 141, 156,
      170, 178, 190, 206, 218, 230,
      260, 268, 280, 289, 296
   Lecture 117, 124, 187, 210, 226, 292
   Recital 38, 59, 93, 104, 112, 124, 141,
      148, 164, 174, 182, 194, 202, 214,
      222, 226, 234, 260, 280, 288
   Rehearsal 59, 66, 96, 132, 249,
      293

Isolation Booth 63, 157

Lobby 42, 59, 78, 132, 152, 161, 174, 186,
214, 248

Opera House 145, 175

Organ Spaces:
   Performance 140, 152, 206, 218,
      230, 260, 288
   Recital 140, 186, 206, 218, 230,
      260, 280, 288

Piano Lab 62, 101, 165, 198, 210

Practice Rooms:
   Band 35, 50, 58, 100, 112, 141,
      148, 174, 210, 242
   Choral 35, 50, 74, 88, 116, 125, 141,
      148, 170, 210, 284
   Dance 35, 42, 62, 116, 165, 206,
      214, 226, 248, 284

Practice Rooms (cont.):
   Ensemble 35, 47, 58, 88, 96, 116,
      141, 170, 198, 206, 231, 239, 261,
      277, 292
   Instrumental 35, 88, 97, 104, 165,
      206, 230, 239, 296
   Jazz 97, 141, 174, 195, 235
   Opera 231, 242
   Orchestra 47, 58, 74, 93, 125, 148,
      156, 165, 174, 182, 235, 243, 252,
      276, 296
   Organ 141, 206, 218, 230, 256
   Piano 62, 92, 104, 116, 141, 165,
      195, 210, 230, 256, 277
   Percussion 35, 88, 97, 116, 141, 174,
      198, 211, 231, 243, 256
   Vocal 104, 211, 230, 284

Recording Studio 42, 54, 62, 82, 93, 104,
152, 165, 183, 214

Rehearsal Rooms/Rooms/Studios:
   Band 43, 54, 62, 75, 88, 100, 112,
      128, 136, 148, 161, 174, 211, 239
   Choral 54, 62, 70, 88, 101, 116, 124,
      136, 148, 156, 170, 178, 194, 206,
      238, 269, 284, 293
   Dance 54, 62, 70, 78, 116, 140, 156,
      165, 206, 214, 227, 248, 284
   Ensemble 38, 55, 62, 70, 82, 96,
      109, 116, 132, 141, 148, 170, 178,
      198, 206, 215, 231, 239, 252,
      260, 269, 277, 288, 296
   Instrumental 70, 88, 97, 104, 117,
      165, 178, 206, 230, 239, 296
   Jazz 54, 67, 97, 132, 141, 156, 174,
      183, 195, 235, 260
   Opera 148, 194, 231, 242, 281, 284
   Orchestra 54, 62, 74, 93, 101, 109,
      124, 133, 148, 165, 174, 182, 215,
      234, 243, 252, 260, 276, 288,
      296
   Organ 140, 152, 185, 206, 218,
      230, 239, 257, 280, 288

Rehearsal Rooms/Rooms/Studios (cont.):

Piano 38, 62, 92, 101, 116, 141, 160, 195, 207, 230, 260, 277

Percussion 51, 63, 88, 97, 116, 141, 174, 198, 211, 231, 243, 260, 272, 289

Vocal 63, 83, 104, 117, 128, 137, 160, 210, 230, 284

String Quartet 141, 215, 272

Teaching Room / Teaching Studio 58, 96, 100, 148, 170, 187, 195, 206, 214, 235, 242, 252, 260, 268, 272, 280, 284, 292

Theatre Spaces:

Black Box 42, 62, 70, 96, 116, 141, 194, 215, 227

Dance 140

Multi-Use 35, 47, 50, 63, 71, 79, 101, 113, 125, 144, 214, 238, 285

INDEX B:
BY LOCATION

Australia

Victoria

University of Melbourne | The Ian Potter Southbank Centre, Southbank — 288

Canada

Alberta

Mount Royal University | Bella Concert Hall, Calgary — 268

Germany

University of Music Karlsruhe | Campus One, Karlsruhe — 144

Singapore

Yong Siew Toh Conservatory of Music | Yong Siew Toh Conservatory of Music Building — 296

Switzerland

Musikinsel Rheinau, Rheinau — 276

United States of America

California

Music Academy of the West | Hind Hall, Santa Barbara — 272

San Francisco Conservatory of Music, San Francisco — 280

Scripps College | Performing Arts Center, Claremont — 124

Colorado

University of CO at Colorado Springs | Ent Center for the Arts, Colorado Springs — 214

Connecticut

Eastern Connecticut State University | The Fine Arts Instructional Center, Willimantic — 178

Western Connecticut State University | Visual and Performing Arts Center, Danbury — 156

Yale School of Music | Adams Center for Musical Arts, New Haven — 242

Florida

Academy of the Holy Names, Tampa — 34

Berkeley Preparatory School | Gries Center for the Arts and Sciences, Tampa — 38

Bethune-Cookman University | Julia E. Robinson Memorial Music Hall, Daytona Beach — 88

Dr. Phillips High School, Orlando — 46

Edgewater High School, Orlando — 50

Hillsborough Community College | Performing Arts Building, Tampa — 100

Southeastern University | College of Arts & Media, Lakeland — 210

70   Trinity Preparatory School, Winter Park

128   University of Florida | Music Building, Gainesville

132   University of Florida | Steinbrenner Band Hall, Gainesville

Georgia

62   Rainey-McCullers School of the Arts, Columbus

Illinois

174   DePaul University | Holtschneider Performance Center, Chicago

54   New Trier Township High School, Winnetka

108   North Central College | Fine Arts Center, Naperville

194   Northwestern University | Ryan Center for the Musical Arts, Evanston

292   Wheaton College | The Armerding Center for Music and the Arts, Wheaton

Indiana

96   Earlham College | Center for the Visual and Performing Arts, Richmond

Iowa

218   The University of Iowa | Voxman Music Building, Iowa City

Kansas

136   The University of Kansas | Murphy Hall: Choral Rehearsal Room, Lawrence

222   The University of Kansas | Murphy Hall: Swarthout Recital Hall, Lawrence

Maryland

140   University of Maryland | The Clarice Smith Performing Arts Center, College Park

Massachusetts

248   Boston Conservatory at Berklee | Richard Ortner Studio Building, Boston

42   Deerfield Academy | The Hess Center for the Arts, Deerfield

260   Indian Hill | Music Center, Littleton

226   University of Massachusetts Boston | University Hall, Boston

78   The Winsor School | The Lubin-O'Donnell Center for the Performing Arts, Athletics, and Wellness, Boston

82   ZUMIX, East Boston

Michigan

182   Hope College | Jack H. Miller Center for Musical Arts, Holland

Minnesota

170   Carleton College | Weitz Center for Creativity, Northfield

Mississippi

104   Mississippi University for Women | Poindexter Hall, Columbus

Missouri

186   Missouri State University | Ellis Hall, Springfield

Nebraska

230   University of Nebraska Omaha | Janet A. and Willis S. Strauss Performing Arts Center, Omaha

New Hampshire

58   Phillips Exeter Academy | Forrestal-Bowld Music Center, Exeter

New Jersey

160   Westminster Choir College | Marion Buckelew Cullen Center, Princeton

New York

264   Manhattan School of Music, New York

190   Nazareth College | Jane and Laurence Glazer Music Performance Center, Rochester

120   Rensselaer Polytechnic Institute | EMPAC, Troy

Ohio

284   University of Cincinnati | College Conservatory of Music, Cincinnati

164   Xavier University | Edgecliff Hall, Cincinnati

Oklahoma

74   Union Public Schools | Fine Arts Building, Tulsa

Oregon

116   Reed College | Performing Arts Building, Portland

234   The University of Oregon | Berwick Hall, Eugene

Pennsylvania

252   Curtis Institute of Music | Lenfest Hall, Philadelphia

256   Curtis Institute of Music | Wyncote Organ Studio, Philadelphia

202   The Pennsylvania State University | Music Building I, University Park

Rhode Island

66   RI Philharmonic Music School | The Carter Center, East Providence

South Dakota

206   South Dakota State University | The Oscar Larson Performing Arts Center, Brookings

Tennessee

148   The University of Tennessee
       Knoxville | Natalie L. Haslam
       Music Center, Knoxville

Washington

198   Olympic College | College
       Instruction Center, Bremerton

112   Peninsula College | Maier Hall,
       Port Angeles

152   Wenatchee Valley College | Music
       and Art Center, Wenatchee

238   Whitworth University | Cowles
       Music Center, Spokane

Wyoming

92   Casper College | Music Building,
       Casper

L. M. Ronsse et al. (eds.), Rooms for the Learned Musician

Acentech Incorporated

248   Boston Conservatory at Berklee | Richard Ortner Studio Building

42   Deerfield Academy | The Hess Center for the Arts

226   University of Massachusetts Boston | University Hall

160   Westminster Choir College | Marion Buckelew Cullen Center

82   ZUMIX

Acoustic Distinctions, Inc.

170   Carleton College | Weitz Center for Creativity

182   Hope College | Jack H. Miller Center for Musical Arts

190   Nazareth College | Jane and Laurence Glazer Music Performance Center

206   South Dakota State University | The Oscar Larson Performing Arts Center

148   The University of Tennessee Knoxville | Natalie L. Haslam Music Center

Acoustics by JW Mooney

164   Xavier University | Edgecliff Hall

Acoustonica, LLC

74   Union Public Schools | Fine Arts Building

Cavanaugh Tocci Associates, Inc.

66   RI Philharmonic Music School | The Carter Center

Clayton Acoustics Group

256   Curtis Institute of Music | Wyncote Organ Studio

D.L. Adams Associates

92   Casper College | Music Building

Jaffe Holden Acoustics

264   Manhattan School of Music

214   University of CO at Colorado Springs | Ent Center for the Arts

218   The University of Iowa | Voxman Music Building

156   Western Connecticut State University | Visual and Performing Arts Center

Kahle Acoustics

276   Musikinsel Rheinau

144   University of Music Karlsruhe | Campus One

296   Yong Siew Toh Conservatory of Music | Yong Siew Toh Conservatory of Music Building

Kirkegaard

252   Curtis Institute of Music | Lenfest Hall

174   DePaul University | Holtschneider Performance Center

178   Eastern Connecticut State University | The Fine Arts Instructional Center

186   Missouri State University | Ellis Hall

194   Northwestern University | Ryan Center for the Musical Arts

58   Phillips Exeter Academy | Forrestal-Bowld Music Center

120   Rensselaer Polytechnic Institute | EMPAC

280   San Francisco Conservatory of Music

284   University of Cincinnati | College Conservatory of Music

140   University of Maryland | The Clarice Smith Performing Arts Center

234   The University of Oregon | Berwick Hall

238   Whitworth University | Cowles Music Center

242   Yale School of Music | Adams Center for Musical Arts

Marshall Day Acoustics

288   University of Melbourne | The Ian Potter Southbank Centre

McKay Conant Hoover Inc.

272   Music Academy of the West | Hind Hall

124   Scripps College | Performing Arts Center

R. C. Coffeen Consultant in Acoustics

222   The University of Kansas | Murphy Hall: Swarthout Recital Hall

Roland, Woolworth & Associates, LLC

104   Mississippi University for Women | Poindexter Hall

Siebein Associates, Inc.

34   Academy of the Holy Names

38   Berkeley Preparatory School | Gries Center for the Arts and Sciences

88   Bethune-Cookman University | Julia E. Robinson Memorial Music Hall

46   Dr. Phillips High School

50   Edgewater High School

100   Hillsborough Community College | Performing Arts Building

62   Rainey-McCullers School of the Arts

210   Southeastern University | College of Arts & Media

70   Trinity Preparatory School

128   University of Florida | Music
        Building

132   University of Florida | Steinbrenner
        Band Hall

Sparling, a Stantec Company

152   Wenatchee Valley College | Music
        and Art Center

Stantec Consulting Services

198   Olympic College | College
        Instruction Center

112   Peninsula College | Maier Hall

116   Reed College | Performing Arts
        Building

TALASKE | sound thinking

268   Mount Royal University | Bella
        Concert Hall

108   North Central College | Fine Arts
        Center

202   The Pennsylvania State University |
        Music Building I

Threshold Acoustics LLC

96   Earlham College | Center for the
        Visual and Performing Arts

54   New Trier Township High School

230   University of Nebraska Omaha
        | Janet A. and Willis S. Strauss
        Performing Arts Center

292   Wheaton College | The Armerding
        Center for Music and the Arts

78   The Winsor School | The Lubin-
        O'Donnell Center for the
        Performing Arts, Athletics, and
        Wellness

Threshold Acoustics LLC / LKAcoustics Design

260   Indian Hill | Music Center

The University of Kansas School of Architecture Acoustics Studio

136   The University of Kansas | Murphy
        Hall: Choral Rehearsal Room

Printed in the United States
by Baker & Taylor Publisher Services